Praise for

AGE OF DANGER

"In the years after the 9/11 attacks terrorism became the zoom-like focus of our government and military. Hoehn and Shanker make a powerful case that our national security leadership requires a more panoramic definition of what is a threat to the United States. Russia and China are back in view but other problems less so. National security must include food security, climate security, and disease security. Hoehn and Shanker clearly define a new outlook and the new set of institutional tools to manage the Age of Danger in which we find ourselves today."

—Chuck Hagel, former Secretary of Defense,
former US Senator, Vietnam veteran

"*Age of Danger* makes a compelling case that we need to re-architect our national security processes and institutions to deal with the challenges of this new era, from great power competition to climate change and pandemics. Creative and thought-provoking, this book is a must read for students, policy practitioners, and concerned citizens alike."

—Michele Flournoy, former Under
Secretary of Defense for Policy

"Tomorrow's threats are likely to include Great Power competition, cyber, disease, and climate—and we are far from prepared to meet them. In this timely volume, two leading experts help us think through new approaches to tune up the vast machine of national security to make ourselves more secure. Time is of the essence!"

—Admiral James Stavridis, former NATO
Supreme Allied Commander and author of
2034: A Novel of the Next World War

"Admiral Bill Crowe used to say 'At times like this, it's important to remember there have always been times like this.' That might have been true in the Admiral's day, but, as Hoehn and Shanker point out in their gripping and persuasive book, we are now living in unprecedented times. *Age of Danger* makes a clear, rational, and urgent case for a significant reevaluation of our national security strategy."

—Admiral Timothy J. Keating, former Commander,
US Pacific Command and US Northern Command

"Our national security structures were built more than seventy years ago. They served us well over time, but, like an old car, there is only so much tinkering you can do. It's time to put the old Chevy in the garage and build a modern national security machine. Hoehn and Shanker offer solutions on how to do it."

—Nadia Schadlow, former Deputy National
Security Advisor for Strategy

"Andy Hoehn and Thom Shanker provide an urgent wake-up call that the range of national security challenges facing the United States is both expanding and growing evermore dangerous—and that we are not adequately prepared to address them. Until the US national security apparatus recognizes that "the future needs a seat at the table," the United States and its people will face increasingly grave dangers from these new and deeply underestimated threats."

—Nora Bensahel, Visiting Professor of Strategic Studies,
The Johns Hopkins School of Advanced International
Studies, and coauthor of *Adaptation Under Fire:
How Militaries Change in Wartime*

"*Age of Danger* leaves us with no excuses. The new, serious threats it describes demand our attention and, more importantly, our action. We are out of time for delay. This is a clear, direct, and understandable must-read for anyone concerned about the nation's security."

—William "Mac" Thornberry, former chairman of
the House Armed Services Committee

Age of Danger

Age of Danger

Keeping America Safe in an Era of New Superpowers, New Weapons, and New Threats

Andrew Hoehn & Thom Shanker

hachette BOOKS

NEW YORK

Hachette Books
Hachette Book Group
1290 Avenue of the Americas
New York, NY 10104
HachetteBooks.com
Twitter.com/HachetteBooks
Instagram.com/HachetteBooks

First Edition: May 2023

Published by Hachette Books, an imprint of Perseus Books, LLC, a subsidiary of Hachette Book Group, Inc. The Hachette Books name and logo is a trademark of the Hachette Book Group.

Library of Congress Cataloging-in-Publication Data

Names: Hoehn, Andrew R., author. | Shanker, Thom, author.
Title: Age of danger: keeping America safe in an era of new superpowers, new weapons, and new threats / Andrew Hoehn and Thom Shanker.
Description: First edition. | New York, NY: Hachette Books, Hachette Book Group, 2023. | Includes bibliographical references and index.
Identifiers: LCCN 2022051322 | ISBN 9780306829109 (hardcover) | ISBN 9780306829116 (trade paperback) | ISBN 9780306829123 (ebook)
Subjects: LCSH: National security—United States. | United States—Military policy. | United States—Foreign relations—21st century.
Classification: LCC UA23 .H5343 2023 | DDC 355/.033573—dc23/eng/20230206
LC record available at https://lccn.loc.gov/2022051322

ISBNs: 9780306829109 (hardcover); 9780306829123 (ebook)

Printed in the United States of America

LSC-C

Printing 1, 2023

CONTENTS

PART I
HOW WE GOT TO THE AGE OF DANGER

CHAPTER 1

INTRODUCTION

American taxpayers are asked each year to open their wallets and pay for a government national security machine that costs more than $1 trillion—yes, *trillion*—to operate, and more than $1.25 trillion if we count the Veterans Administration.[1] Just for one year. That bill includes money for the military, intelligence community, homeland security, diplomacy, federal law enforcement, and salaries for all the elected officials, political appointees, diplomats, and civil servants who work to defend the United States at home and abroad.

Given this annual $1.25 trillion price tag, we have two questions: How is it possible that our public institutions seem unprepared and get it wrong on so many critical issues so often? And, if our fabulously expensive government national security machine is not working, what is to be done?

It is often said that generals are best at preparing for the last wars. We are close enough to the generals (and admirals) to know that this is an exaggeration, but not entirely untrue, when it comes to the institutions they serve. The National Security Act of 1947, designed after World War II, gave us the basic

[1] Based on Congressional Budget Office estimates for FY2023.

system we still use today to determine threats to the United States and how to respond to them. It served the country well—mostly well—during the Cold War because it was set to detect and respond to a panorama of global threats. It was adapted after the Cold War—and substantially overhauled after 9/11.

But that most recent overhaul, following the attacks of September 11, 2001, retooled the national security machine to focus mainly on one major threat—terrorism. We now realize, painfully well, that a zoom-like attention on al-Qaeda, its affiliates, and its rivals came at the cost of tuning out the vast array of new threats we now confront, some of them surprises. To be sure, any deaths from terrorism above zero are a tragedy, but the Bush administration erred in its rallying cry calling al-Qaeda a threat to the very existence of the United States on a par with, say, Russia's nuclear arsenal. It is notable that Michael Leiter, who directed the National Counterterrorism Center, a post-9/11 institution, believes the nation went too far in defining terrorism as an *existential* threat to the United States, one that overshadows other risks.

"When we killed bin Laden, I knew it [terrorism] was declining strategically," Leiter said.[2] "But even before then, I saw it as of declining strategic importance to the United States compared to cyber, the rise of China, a resurgent Russia. It was clear to me that we've probably overreacted some to terrorism."

Even so, the threat of terrorism has not vanished, and it would be foolish to declare total victory and then forget about it. The July 2022 killing of al-Qaeda leader Ayman al-Zawahiri demonstrated the importance of sustained vigilance. The nation's

[2] Michael Leiter (former director of National Counterterrorism Center), interview with authors, November 2021.

law enforcement agencies, homeland security forces, intelligence community, and the military seem to have arrived at the more balanced assessment that, if terrorism is a problem that cannot be eradicated, it can be managed without being an obsession that blinds our leaders to other threats.

The point of this book is to showcase how current and emerging threats are not being spotted early enough or, if they are spotted, are not warned with sufficient clarity and urgency to prompt action. Or if they are spotted and acknowledged, the machinery might not be in place to deal with them. When people say, "Didn't the government know about this?" the question really is "Who knew about this? At what level of authority? And what else was going on?"

A truism of government leadership is that the inbox is always overflowing for the president and all the cabinet-level officials responsible for keeping our country safe. How do they set priorities?

If US strategy over the last seventy-five years could be summed up as a maxim, it would be make friends and keep threats at a distance. That was the basic lesson we drew from fighting two world wars. That is what guided US strategy during the Cold War. It led to alliances with the other industrial democracies—and occasionally with dictatorships whose interests aligned in pragmatic ways. And it's what motivated George W. Bush and his administration in the aftermath of the 9/11 attacks.

Today, with terrorism fading from its two-decade role as Public Enemy No. 1, a host of new challenges, and of a different kind, are rising. Threats used to manifest in the form of bombs and bullets, whether from a nation-state, insurgents, or terrorists. But as the wars in Afghanistan and Iraq wound down, attention indeed did turn to the threats from digits, storms, and germs. Not one of those threats can be kept at a distance. Oceans and border walls cannot hold them out. A specific mechanism for acting against them does not exist. To be sure, COVID-19 prompted the United States to redefine national security away from problems that can only be solved by an overextended military—which nonetheless played a role even in pandemic response—and to elevate other mechanisms to save lives and secure the nation.

But then Putin invaded Ukraine, and the United States was facing the worst of both worlds: a hostile, nuclear-armed rival stalking its neighbor's sovereign territory and undermining global order while the United States was still battling a pandemic and an increase in cybercrime that shut down a national pipeline and heightened concern over outright cyber war with the Kremlin. Western attention had not focused sufficiently on Putin, with the US national security establishment ridiculing Russia, a former superpower, as irrelevant, an Upper Volta with missiles. But a dying man with a gun can kill you just as dead as a healthy man with a gun. It took a Russian army on the Ukrainian border, and then invading, to wake the government up to the threat posed by Putin and his compatriots.

These examples suggest that the US government remains challenged to answer even a most basic question: What is national security? How do we define true mortal, even existential, threats? What does it mean to keep our country safe? Is success measured in lives saved versus lives lost, amount of territory protected from an invader, or damage to our economy averted, or all of the above?

Just under three thousand people died on 9/11, the event that launched twenty years of war with an estimated price tag of $6.5 trillion.[3] Compare that to COVID, which has claimed vastly more American lives—one million and counting—and produced an economic recession. So, what does it mean to safeguard American lives and interests in the modern era? If the government can't even answer that, then it can't figure out the best way to keep us safe.

The warnings over "cyber war" provide a perfect example.

If this was 2012, some would say the most significant threat is a "cyber Pearl Harbor"—crippling the electrical grid, water supply, or banking system—routinely warned by senior officials. If this was 2014, some would say that the most significant cyber threat was North Korea hacking into corporate computers, such as Sony. If this was 2016, then some would cite Russian hacking to disrupt American elections and undermine our democracy. And if this was 2021, the biggest threat perceived would be cybercriminals, likely abetted by Russia, using ransomware to shut down a critical pipeline, sending gas prices spiraling up and the economy spiraling in the opposite direction.

Or maybe the "cyber Pearl Harbor" sneak attack has already happened, not as a lightning strike but as the slow and patient secret infiltration by China of private-sector computer systems, including of defense contractors, to steal intellectual property worth billions of dollars, a move that likely has saved China years of military research and development.

---

[3] This reflects a higher-end estimate of costs, though not the highest estimate. Some analysts dispute including Department of Homeland Security spending in the estimate. Neta C. Crawford, "United States Budgetary Costs and Obligations of Post-9/11 Wars Through FY2020: $6.4 Trillion" (research paper, 20 Years of War: A Costs of War Series, Watson Institute at Brown University and Frederick S. Pardee Center at Boston University, November 13, 2019), https://tinyurl.com/yjra47z4.

Which is it? Or something else? Maybe talking about cyber war as war and combat only makes us less safe. After all, if fuzzy analysis leads you to define every problem as a nail, then only a hammer is the perfect tool. And it's not.

Other major threats that are not receiving enough attention right now include armed drones, climate change as a destabilizing geopolitical event, and pandemics that affect not only humans but livestock and crops. Shouldn't food security be part of national security? A catastrophic crop blight—whether emanating from nature or a terrorist's test tube—could cause mass starvation and cripple the nation's economy. We will take you to the laboratory in the American agricultural heartland where researchers say we already may be too late in preparing for this threat.

Drones were once an American military monopoly. Now adversaries are fielding off-the-shelf varieties, outfitting these small, hard-to-detect aerial machines to fly over cities carrying weapons and explosives and, potentially, chemical and biological weapons. Much attention has been given to Iran's nuclear weapons program, but it was Iran's use of drones that brought Saudi Arabia's oil production to a halt in September 2019. It was to Iran that Russia turned when it needed drones to fight its war in Ukraine. The image of Vladimir Putin kowtowing to Iran's leadership is not something most observers would have expected. Except Iran had the drones that Putin needed.

Here at home, as the counting of ballots in the 2020 presidential election indicated Joe Biden would win, the Federal Aviation Administration ordered a total ban on drones over his residence in Delaware, warning that any unmanned aircraft in the security zone would be destroyed by the Department of Homeland Security or military forces, proving that the US government fully recognized the domestic drone threat.

And though most responsible citizens and leaders acknowledge the risk of climate change, few are focusing on other significant effects that could be just as catastrophic as rising temperatures: mass migration and mass starvation that could further destabilize fragile nations around the globe. The Pentagon, which spends billions on creating "force multipliers," is only belatedly understanding how climate change is a "threat multiplier." The nation's most important coastal installations for warships and Marines and combat jets could be under water, and useless, with rising sea levels. Increasing heat and moisture in the atmosphere affect the ability of military helicopters and warplanes to take off. That is a double curse because it will require aircraft to burn even more fossil fuels to meet the increasing need for lift. Instability from climate change can be seen abroad in such heartbreaking examples as the brutal civil war in Syria, at least partially sparked by mass migration to the cities owing to long-term drought.

Experts caution that even as the world sets to battle back against the COVID-19 virus, another pandemic is certain—with monkeypox already a threat in mid-2022. Will the nation have learned the lessons—and, more importantly, incorporated those lessons into national security planning? Are our government institutions up to the task?

What is to be done?

This country's national security system needs an overhaul, a retooling that rivals the major changes made at other critical turning points in history: the end of World War II, after the collapse of the Soviet Empire in 1989, and post-9/11.

The United States is tired of war. But it must find the energy, creativity—and money—to create an industrial-strength *life-saving machine* as a counterpart to a system that, for too long, was focused on being the best at threatening lethal force to deter adversaries and carry out military operations if—when—deterrence failed. A lifesaving machine is needed to protect American lives from digits and microbes as much as from planes, bombs, and bullets.

In this book, using exclusive research, we focus on what is wrong and how to fix it by breaking down the $1.25 trillion national security apparatus into two major pieces.

The first piece we call "the warning machine." Based inside the intelligence community, but significantly involving the Pentagon and the Department of Homeland Security (itself created as a response to 9/11), intelligence professionals gather, curate, and assess information from around the world 24/7/365. The many parts of the warning machine constantly push information and warnings to the vast national security apparatus, including a daily brief to the president himself.

The warning machine is generally good at watching and sizing up problems. But it can't watch everything. Sometimes the warning machine misses threats, even big threats like 9/11. Sometimes it observes a threat but does not correctly measure its immediacy or impact. And sometimes it warns perfectly... but the warning fails to prompt the required response in the other half of the system, which we call "the action machine." This machine consists of a large and expensive set of systems, from the FBI and Homeland Security to the military, charged with taking the actions to keep our country safe.

We should be clear. When we talk about the machines, we mean the many people, processes, organizations, and supporting technology and infrastructure that are responsible for keeping the country safe. The machines are elaborate in design and have enormous scope and responsibility. They encompass everyone from the president to the private or clerk responsible for carrying out a particular task. The people are assigned across the federal workforce, the private sector that supports the federal workforce, and, increasingly, states and localities. We refer to the machines as a way to simplify our descriptions. They are anything but simple.

In the chapters ahead, we analyze times when officials indeed heard an accurate warning with sufficient clarity and alarm. And we examine when and why they did not. We also consider what other pressures and pressing issues—the tyranny of limited resources and of measured political capital—prevent the action machine from making appropriate decisions and carrying them out. We include cases of when the machine itself didn't have the right tools to get the job done, though, as some shrewd observers note, getting the job partly done is better than not getting the job done at all.

The grimmest decision any democracy can make is the decision to go to war. The threat of invading conventional armies? We can fight them—with industrial precision. North Korea preparing for a missile test? We are carefully watching that one, too. But preparing for the next war if it's a new kind of war? Putin ordering cyberattacks on the integrity of our voting system or China taking advantage of the two decades the United States spent circling the drain in the strategic cul-de-sac of Iraq to make gains on us, or even overtake us? Missed them or thought we had more time than we actually did. The rise of ISIS? Response came only after a region the size of Britain in northern Iraq and Syria was nearly overrun. Pandemic disease? Many saw it coming, but most of the response pieces were not in place. The effects of climate change on the American military's ability to operate effectively? Barely on the public radar.

Some of these threats are over the horizon, and some are just around the corner. Will the warning machine notice them in time and issue a clear call to action to both prevent the next crisis and goad the action machine into taking the appropriate steps if a threat reaches our shores? And will the action machine respond quickly and forcefully enough? Is the full lifesaving machine even built?

The national security machine cannot shed its prowess for dealing with threats like those emanating from Russia today or perhaps China tomorrow and focus only on new age threats. We need a machine capable not of a biathlon or triathlon but of a decathlon. We will never be able to do everything, and strategy and statecraft involve the art of making choices. But in building the machine needed to make America secure, we need to keep the lens on wide-angle, and we need the tools to act across a range of actual threats, both old and new. It is clear that, since 9/11, the national security machine has focused on the "today" threat. As we move past the Forever Wars to face new threats, it is even clearer that the future needs a seat at the table.

"We sometimes act as if all of this is just so easy," said Eric Edelman,[4] a four-decade pillar of American national security as the Defense Department's undersecretary for policy—considered the no. 3 Pentagon post—as well as ambassador to Turkey and Finland.

"There are a lot of historians who look back at all this and say, 'Well, they should have known this and they should have known that,' " Edelman added. In response to such claims, he

[4] Eric Edelman (former undersecretary for policy at the Defense Department), interview with authors, October 2021.

cites the work of a British historian, Ian Kershaw, who defined the challenge this way: history is lived forward, but it is only understood backward.

"The reason we get things wrong so much in national security," Edelman concluded, "is because it's really fucking hard."

CHAPTER 2

THE WARNING MACHINE AND THE ACTION MACHINE

For the first 170 years of its history, the United States operated with a foreign policy machinery built for war and peace, without a whole lot in between. That is not to say the United States only operated in the realms of war and peace. When President Thomas Jefferson decided it was bad policy to continue paying tributes to the Barbary pirates operating off the shores of North Africa, he dispatched the US Marines to bring an end to this particular reign of terror and ransom. Jefferson did not declare war in dispatching Marines but rather called upon the Marines as a tool of foreign policy. This piece of history is captured in the opening line of the Marine Hymn: "from the halls of Montezuma to the shores of Tripoli." Jefferson was choosing a path between war and peace, but there was little organizational machinery then between the president and the Marines aboard Navy ships who would implement his decisions. The modern warning and action machines would come much later.

Indeed, for more than a century and a half, the United States national security machinery comprised the Departments of State,

Treasury, War, and Navy. There were coordinating mechanisms within the federal government, to be sure, but no standing bodies or committees organized to support the president, nothing akin to the modern National Security Council system. There also was no standing intelligence function—no warning machine—to provide notice of pending threats other than what was produced through dispatches from foreign embassies and military deployments and communications from private citizens. Prior to the telegraph, news and information from abroad traveled at the speed of a sailing ship and, later, a coal-fired vessel. That meant news traveled slowly and those acting on instructions from the president had considerable autonomy. Undersea cables, which appeared in the late 1850s, allowed for more instantaneous communications over long distances, but they did not initially provide widespread coverage. Transatlantic telephone lines only began operations in the 1940s.

The first instance of reports from the battlefield being delivered to leaders and readers in something near real time was during the Civil War, at Gettysburg, when newly installed telegraph lines allowed rapid dispatches from the front. These missives were read immediately by President Lincoln, who crossed from the White House to the ticker inside his nearby War Office, and they landed on the front page of newspapers across the industrialized Northern states within hours. It was the Victorian age equivalent of CNN's "Bombs are falling on Baghdad."[1]

World War I and its aftermath would have put this loosely knit structure of diplomatic and military instruments to a test

---

[1] "The Times at Gettysburg, July 1863: A Reporter's Civil War Heartbreak," *New York Times*, July 4, 2018, https://www.nytimes.com/2018/07/04/insider/the-times-at-gettysburg-july-1863-a-reporters-civil-war-heartbreak.html.

had Woodrow Wilson's ambition for the League of Nations survived a vote by the US Senate. Wilson envisaged a large-scale role for the United States in the planned new league and negotiated mechanisms for US leadership. The president spent much of the first half of 1919 negotiating the peace talks abroad, including visits to Paris, London, Rome, and Brussels.[2] He imagined an activist policy of involvement and intervention through the mechanisms of the league he helped form. Wilson would be awarded the Nobel Peace Prize for his efforts. But concern that the treaty might commit the United States to defending other league members should they come under attack led the Senate to reject Wilson's efforts by a substantial margin.[3] As it was, the United States chose isolation—for the most part—over entanglement, and its institutional mechanisms—the predecessors to the modern warning and action machines—would not be tested again in a serious way until World War II.

The lasting memory of triumph in World War II—the heroic work of the Greatest Generation—tends to cloud how fraught the decision-making process was during the war. Some of this had to do with President Franklin Roosevelt himself. Roosevelt was notorious for cultivating multiple layers of decision processes, making different promises to different people, and keeping even some of his most important advisers in the dark on critical matters. Henry Stimson, Roosevelt's secretary of war, expressed in his diary his own dismay: "The president is the poorest administrator I have ever worked under in respect

---

[2] As noted by the State Department Office of the Historian, https://history.state.gov/departmenthistory/travels/president/wilson-woodrow.

[3] United States Senate, "The Senate and the League of Nations by Henry Cabot Lodge (1925)," https://www.senate.gov/reference/reference_item/Versailles.htm.

to the orderly procedure and routine of his performance. He is not a good chooser of men and he does not know how to use them in coordination."[4] Tough words from one of Roosevelt's closest advisers.

But it was not just Roosevelt, as complicating as his behavior could be. The machinery itself proved to be antiquated. It was an eighteenth-century design being used for twentieth-century problems. Intelligence flowed through separate organizational stovepipes without standing coordinating mechanisms. Japan's December 1941 attack on Pearl Harbor, which in retrospect had been signaled in so many different ways, proved to be the single greatest failure of warning in national security to date in US history.

Roberta Wohlstetter wrote the landmark book on the intelligence failure that led to the attack. Her conclusions sound eerily similar to the assessments that followed the 9/11 attacks on New York and Washington:

> A curious kind of numbness seemed to characterize these last moments of waiting, a numbness that was an understandable consequence of long association with signals of mounting danger. The signal picture had been increasingly ominous for some time, and now apparently it added up to something big, but not very definite.... There was also a fundamental passivity connected with the avowed policy that the United States could not strike the first blow.[5]

---

[4] Charles A. Stevenson, "The Story Behind the National Security Act," *Military Review*, May–June 2008.

[5] Roberta Wohlstetter, *Pearl Harbor: Warning and Decision* (Stanford, CA: Stanford University Press, 1962), 277.

Wohlstetter went on to conclude:

> After the event, of course, a signal is always crystal clear; we can now see what disaster it was signaling since the disaster has occurred. Before the event, it is obscure and pregnant with conflicting meanings.[6]

This sentiment, of course, would be repeated in the decades that followed.

As the war unfolded, military planning was hampered by interservice squabbles and the lack of formal coordinating mechanisms. The Army, Navy, and Marine Corps were the three standing military services. The Army, along with its aviation corps, reported to the secretary of war, who reported to the president. The Navy, with its own aviation arm, and the Marine Corps reported to the secretary of the Navy, who reported to the president. The War and Navy Departments had coordinating mechanisms, including the Joint Chiefs of Staff, but US military operations still ran through two cabinet-level departments that reported directly to the president. Neither cabinet department was required to coordinate with the secretary of state, who ostensibly was responsible for US foreign policy on behalf of the president. The Army and Navy operated in largely separate spheres throughout the war. A small artifact of history worth remembering is that General Douglas MacArthur operated his Army headquarters out of Australia after he evacuated from the Philippines. Admiral Chester Nimitz operated the Navy headquarters from Hawaii. The Coast Guard operated under the direction of the US Treasury until it was

---

[6] Wohlstetter, *Pearl Harbor*, 387.

temporarily assigned to the Navy in 1941, only to be returned to the US Treasury in 1946.

In his 1951 Pulitzer Prize–winning novel *The Caine Mutiny*, author Herman Wouk described the Navy as an institution designed by geniuses to be run by idiots. Aside from tradition, not a great deal of organizational genius guided the war effort. That is not to take away from individual genius or heroics, but the war effort persuaded most who were involved that a new organizational model needed to be created to guide the United States beyond the war. The United States was victorious in World War II to be sure, but most who were involved in the victory recognized the country was not organized for the world that would follow. In other words, the United States won the war despite the weaknesses of the warning and action machines that were built to support it.

BUILDING THE COLD WAR MACHINE—WIDE-ANGLE VIEW WITH A SOVIET LENS

The National Security Act of 1947—resulting not from crisis and defeat but from victory—provided the architecture for that new national security model. As the United States emerged from World War II, national leaders concluded the United States no longer had the luxury of withdrawing from the world. That was the path chosen at the end of the first great war of the twentieth century, and it led to a second great war a short twenty years later. There was a general view, though far from a consensus, that the United States needed to lead the postwar world that was emerging. No other industrial power was capable of doing so—Germany and Japan had just been defeated, and Great Britain and France had suffered terrible losses during the war. The United States had embraced the Soviet Union as an ally during the war, but it was soon clear the two powers were on diverging paths. Winston Churchill's famous "Iron Curtain" speech signaled the actual breaking point between the Western powers and the new Soviet Empire.[7] America's postwar leaders created a system at home to deal with the emerging world and developed partnerships with the industrial democracies. Many of these alliances, which would prove to be enduring, were pitted against authoritarian regimes that controlled centrally organized economies. The shorthand for this is communist systems, though they came in a variety of forms, the Soviet Union and China being the most prominent examples.

The United States would lead the industrial democracies, and it needed a wide-angle view of the world and the tools to deal with the problems this world would present. It had a lens, of course, to contend with the Soviet threat. The leaders built a machine that had the best engineering and industrial efficiency of the time. They were creating a system for the long haul and were quite expansive in their view of what was required. In a way, we might think of it as a top-of-the-line 1947 Chevy. Not quite a Cadillac, Buick, or Oldsmobile, because Harry Truman was known to be something of a spendthrift.[8] But they built the machine knowing that it had to last. In a few cases they added chrome and shiny hubcaps—think of the modern Air Force or the newly formed CIA. It was built to take the country decades into the future.

---

[7] Winston Churchill, "Sinews of Peace" (speech transcript, Westminster College, Fulton, MO, March 5, 1946), https://www.nationalarchives.gov.uk/wp-content/uploads/2019/05/FO371-51624.jpg.

[8] "Truman Fiscal Aims Assailed as Spendthrift, Peril to U.S.; Wherry-Bridges Attack Comes After Senate Passage of Fund for Ex-GI Aid—White House 'Corrects' President on Deficit Truman Attacked on Fiscal Policies," *New York Times*, April 15, 1949, https://www.nytimes.com/1949/04/15/archives/truman-fiscal-aims-assailed-as-spendthrift-peril-to-u-s.html.

The new warning and action model established a Director of Central Intelligence, who would coordinate intelligence activities, and the Central Intelligence Agency, which had separate directorates for analysis and operations. This marked the beginning of what would become a vast warning machine. The CIA's charge was global, given that the United States had taken on global obligations. Its lens was all things Soviet, wherever they took place, arguably at times even when the Soviet Union wasn't involved. This model created the modern Department of Defense, with four military services—including the brand-new United States Air Force—under a single civilian leader.

Importantly, it also established the National Security Council, or NSC, as a standing structure to advise the president on national security matters. The NSC was to advise and coordinate on policy, not make it. Nor was the NSC designed to oversee implementation of policy. That would happen through the cabinet departments and agencies. The NSC was established under the chairmanship of the president, with the following seven officials as permanent members: the president, the secretaries of State, Defense, Army, Navy, and Air Force, and the chairman of the National Security Resources Board. The president could designate "from time to time" the secretaries of other executive departments and the chairs of the Munitions Board and the Research and Development Board to attend meetings. Although the new Central Intelligence Agency was to report to the NSC, the Director of Central Intelligence was not a member of the council but did attend meetings as an observer and resident adviser.

This basic organizational structure has now been in place for over seventy-five years. But it is worth remembering that all wasn't well at the beginning. Harry Truman, who supported the legislation, was suspicious of the idea of a National

Security Council reporting to the president. He believed it was an impingement on his executive authority and was inching closer to a British-style cabinet government, where the prime minister is the first among equals but not the chief executive. As far as Truman was concerned, the buck stopped with him.

> Despite his service in the Senate, president Truman was a strong defender of presidential prerogatives and responsibilities. He made little use of the NSC before the Korean War, in part because he believed that only the president could make decisions or be accountable. "There is much to this idea," Truman wrote in his memoirs. "In some ways a Cabinet government is more efficient—but under the British system there is a group responsibility of the Cabinet. Under our system the responsibility rests on one man—the president." In the final stages of the legislation, Truman had his staff insist on word changes to make the NSC clearly advisory, with no power to coordinate or integrate policy.[9]

Truman remained dubious of the NSC even as he faced a number of important decisions—the Berlin Airlift, the establishment of the North Atlantic Treaty Organization (NATO), and the Marshall Plan, which helped provide for the economic rescue of Western Europe. It was not until North Korea invaded the south that Truman called upon the NSC as part of the new warning and action machinery.

Unification of the military services was even trickier. Coming out of the war, the War and Navy Departments had compet-

---

[9] Stevenson, "Story Behind the National Security Act."

ing plans. The Army, in particular, advocated for full unification and a streamlining, even specialization, of functions. For example, the Army wanted all land warfare functions to be assigned to the Army, including those provided by the Marines, who were part of the Navy Department. Truman backed the idea of combining the War and Navy Departments. But the Navy, under the leadership of James Forrestal, was suspicious and sought more autonomy for the naval forces, including the Marine Corps. Adding to the challenge were suggestions that the Navy forgo its aviation arm, despite the importance of carrier aviation to the war effort, and that the Marine Corps be limited in numbers or even eliminated.[10] Truman himself did not sit above the debate: "The Marine Corps is the navy's police force and as long as I am president that is what it will remain. They have a propaganda machine that is almost equal to Stalin's."[11]

Regarding the need for reform, Truman lamented, "If the Army and Navy had fought our enemies as hard as they fought each other, the war would have ended much earlier."[12]

Still, the National Security Act of 1947 created the basic architecture for the modern warning and action machines, reflecting the best organizational design of its era. It was machinery that would be operated not just with white shirts and dark ties but also with telephone operators, couriers, clerks, cooks, even soldiers and sailors on KP duty. It was very labor-intensive machinery. It was built to support America's friends and contain a Soviet threat.

---

[10] Stevenson, "Story Behind the National Security Act."

[11] Stevenson, "Story Behind the National Security Act."

[12] David Rothkpof, *Running the World: The Inside Story of the National Security Council and the Architects of American Power* (New York: PublicAffairs, 2006), 51.

Building the machine required building an infrastructure to support it. Military infrastructure to be sure, but also laboratories, training facilities, and schools. Although it's a myth that Eisenhower conceived the modern interstate highway system—it was envisaged a couple decades earlier—he recognized its value in moving troops and equipment if needed and evacuating cities in the event of a nuclear attack. National laboratories, some of which were created to support the war, were maintained at locations like Livermore in California and Los Alamos and Sandia in New Mexico. In time, the modern space infrastructure would follow, with launch facilities in Florida and California and research facilities in Alabama, Louisiana, Maryland, New Mexico, Ohio, Texas, and West Virginia.

Not to mention the vast military infrastructure and supporting industry, including public shipyards. Bases and facilities were concentrated along the two coasts and large military facilities were built across the South and deep into the heartland.

There was also a vast infrastructure overseas. The end of World War II left large troop concentrations in Germany and Japan. The immediate occupation of Germany involved nearly 750,000 Americans, not just troops but also large numbers of civilians, who provided the backbone for the immediate postwar government. The same was true for Japan. The Korean War that soon followed left a large troop presence in South Korea. But the network of bases and infrastructure spread even farther into Iran, Libya, Turkey, and the Philippines. Much of this overseas infrastructure supported the burgeoning warning machine and served as the bulwark of the new containment strategy, where the warning machine would watch and listen, and where the action machine was poised to respond if needed. But for the Philippines, most of the new overseas infrastructure came as a result of war.

Of course, all this created jobs, not just military jobs, but also federal civilian jobs. The federal workforce in the 1950s was 1.9 million, with more than half working for the Defense Department.[13] Jobs flowed through the government and through the vast web of industry contracts. It is easy to think of defense contracts, which are substantial, but the human part of the vast machinery needed housing, clothes, food, and entertainment, even a Coke or Pepsi on occasion. The people who populated the machine wanted the benefits of modern life—homes in the suburbs, supermarkets, washers and dryers, dishwashers, vacuum cleaners, and console TVs, not to mention big, sleek automobiles. They wanted schools, swimming pools, tennis courts, camps, and churches for their kids. The national security machinery required its own machinery to support modern life.

The warning and action machines would be tested and modified over time but would remain remarkably true to the original design. The warning machine was fine-tuned to track every Soviet move using the best available technology and finest human talent. The early space race was less about putting a human on the moon than it was about watching each other's military developments by looking down through cameras in space. Spying itself became a trade with a whole tradecraft built around it. Novels and movies from the likes of John le Carré spun tales around the massive human and technological apparatus. The early engineers of the warning machine knew it would require tinkering. Arthur Vandenberg—identified as Mr. A in order to keep his name secret at the House of Representatives

[13] "Policy, Data, Oversight: Data, Analysis & Documentation," US Office of Personnel Management, https://www.opm.gov/policy-data-oversight/data-analysis-documentation/federal-employment-reports/historical-tables/executive-branch-civilian-employment-since-1940/.

hearing on the new legislation—warned of US inexperience in spying. Vandenberg didn't mince words in his assessment: "We are tyros in the game of foreign intelligence,"[14] meaning the United States would need to experiment and adapt. The growing pains would become evident, not just what was missed or misunderstood, but also, in some regrettable action in places like Guatemala, Iran, and Cuba, what were misguided efforts to replace foreign leaders out of fear of a communist threat.

The action machine was similarly adapted to the needs of the Cold War. Initial attention was devoted to deterring large wars fought with nuclear weapons. The new US Air Force, which required long-range aircraft and missiles to threaten Soviet leaders with nuclear retaliation should the Soviets attack the United States or one of its allies, was very much in its ascendancy. This showed in the budget allocations of the time. Following the Korean War, the Air Force share of the total budget leapt ahead of those of the Army and Navy, at times roughly equaling the share of the other two military departments combined.[15] The massive investment in the Air Force was a reflection of the Eisenhower administration's broader policy of containment backed by the threat of "massive retaliation," as articulated in 1954 by Secretary of State John Foster Dulles. This investment led to a series of

---

[14] From House Hearings, p. 13. https://books.google.com/books?id=2txRifTWSI0C&printsec=frontcover&source=gbs_ge_summary_r&cad=0#v=onepage&q&f=false.

[15] Office of the Under Secretary of Defense (Comptroller), *National Defense Budget Estimates for FY 2022* (Washington, DC: Office of the Under Secretary of Defense, August 2021), https://comptroller.defense.gov/Portals/45/Documents/defbudget/FY2022/FY22_Green_Book.pdf (hereafter, FY22 Green Book). By contrast, in 2022 the Department of the Navy (Navy and Marines) and Department of the Air Force (Air Force and Space Force) received roughly equal funding shares. The Department of the Army share is approximately 20 percent lower.

innovations in modern aircraft, including intercontinental bombers and aerial refueling aircraft. It also led to the development of intercontinental missiles and space orbiting vehicles, later known as satellites. The action machine needed a whole new industry to support it. This gave birth to the modern aerospace industry.

It also led to something of a cat-and-mouse game in the pursuit of technological advantage. Indeed, in 1957 during the Eisenhower administration, Russia successfully placed Sputnik 1 in orbit, which officially marked the beginning of the space race. This represented just one part of the competition that defined the Cold War. References to Sputnik as a shorthand for surprise continue to this day, as we will see later.

A central feature of the new warning and action machines was the creation of alliances. They were to be the bulwark to contain the Soviet Union. The first and most enduring was the establishment of NATO. The alliance was formed around the idea that an attack against one is an attack against all, a concept the US Congress would not embrace when Wilson attempted to make this policy through the League of Nations. The idea is enshrined in Article 5 of the North Atlantic Treaty.[16]

---

[16] Article 5 of the North Atlantic Treaty states:

The Parties agree that an armed attack against one or more of them in Europe or North America shall be considered an attack against them all and consequently they agree that, if such an armed attack occurs, each of them, in exercise of the right of individual or collective self-defence recognised by Article 51 of the Charter of the United Nations, will assist the Party or Parties so attacked by taking forthwith, individually and in concert with the other Parties, such action as it deems necessary, including the use of armed force, to restore and maintain the security of the North Atlantic area.

Any such armed attack and all measures taken as a result thereof shall immediately be reported to the Security Council. Such measures shall be terminated when the Security Council has taken the measures necessary to restore and maintain international peace and security.

NATO, created in April 1949 as the Cold War was intensifying, originally had twelve founding members.[17] Greece and Turkey became members in 1952; the Federal Republic of Germany (West Germany) became a member in 1955; and Spain entered the alliance in 1982. NATO, too, has its own machine, with both political and military structures. The political structure is housed in the North Atlantic Council, which includes permanent representatives or ambassadors from every NATO member. Its job is to facilitate diplomatic interactions across the member states. Of course, the larger the number of members, the trickier the diplomacy. The military structure is guided by the NATO Military Committee, which provides oversight of the military commands and acts as the planning and coordinating function for potential wartime operations. The NATO military commander, Supreme Allied Commander Europe, or SACEUR, is headquartered in Belgium. Dwight D. Eisenhower was the first SACEUR. He established the Supreme Headquarters of Allied Powers Europe (SHAPE) in Mons, Belgium.[18]

NATO added new members after the Cold War. Poland, Hungary, and the Czech Republic joined in 1999; Bulgaria, Estonia, Latvia, Lithuania, Romania, Slovakia, and Slovenia joined in 2004; Albania and Croatia joined in 2009; Montenegro joined in 2017; and North Macedonia joined in 2020. Finland and Sweden sought membership in 2022—a direct result of Russia's war on Ukraine. The United States Senate

---

[17] NATO's original twelve members are Belgium, Canada, Denmark, France, Iceland, Italy, Luxembourg, the Netherlands, Norway, Portugal, the United Kingdom, and the United States.

[18] A lighter piece of NATO history occurred in the early 2000s, when then Secretary of Defense Donald Rumsfeld greeted General James Jones and Admiral Edmund Gambastiani, both NATO Supreme Allied Commanders, as "the Supremes." Hoehn was present for the event.

supported Finland's and Sweden's applications in August 2022; action from other NATO member governments is still pending. While many have wondered about NATO's durability after the Cold War, it has proven to be an especially durable organization. Unlike other alliance relationships, many have wanted to join the alliance and no member has left, though France has signaled a desire on several occasions.

Other important alliances are primarily bilateral. The US–Japan alliance was established in 1951. The ANZUS treaty (Australia, New Zealand, United States) was also established in 1951, as was the treaty with the Philippines. A treaty with South Korea followed in 1953 as the war in Korea was winding down.

The Southeast Asia Treaty Organization, or SEATO, was established between the United States, France, Great Britain, New Zealand, Australia, the Philippines, Thailand, and Pakistan in 1954. This can be thought of as overbuilding the machinery. It was clear at the time and certainly in retrospect that SEATO came in response to Mao's triumph in China and a perceived communist threat to all of Southeast Asia. Still, it represented an odd mix of British Commonwealth countries, former colonial powers (Great Britain and France), and developing nations. The ties that bound the allies were much looser than the forces pulling them apart. Members began to withdraw in the early 1970s. Pakistan was unhappy with the organization's inaction during its war with India, and neither France nor Pakistan supported the US war in Vietnam. The organization formally disbanded in 1977.

The Baghdad Pact, founded in 1955 as an alliance of Turkey, Iraq, Pakistan, Great Britain, and Iran, was clearly intended to contain Soviet expansion to the south and was meant to protect critical access to Middle East oil supplies, the lifeline for the industrial economies. Like SEATO, it was an odd mix of former colonial powers, oil-rich Middle East states, and the newly independent Pakistan. To the United States, which joined the alliance in 1959, and Great Britain, the Baghdad Pact looked like another bulwark against Soviet expansion. To the regional members, the alliance was seen as an opportunity to share in the new industrial wealth and as a hedge against threats from one another, neither of which proved to be forthcoming. The Baghdad Pact was renamed the Central Treaty Organization, or CENTO, after Iraq withdrew in 1959. Following the Iranian revolution and Iran's subsequent withdrawal from the treaty, CENTO was disbanded in 1979.

Indeed, the various mutual defense treaties were all part of the larger Cold War structure. Some proved more resilient than others, with the NATO, Australia, Japan, and South Korea relationships proving to be most resilient. Not surprisingly, the most robust of these relationships included the industrial democracies—in Europe and, in time, across East Asia. All had an important stake in the emerging world order where security—the oxygen, if you will—was necessary to allow for the free flow of goods and services. The industrial democracies had a critical stake in the larger global order.

If security was the oxygen, then fossil fuel energy was the lifeblood. Thus, the importance of the Middle East to the industrial democracies. This significance first became clear during World War II and was captured for history when President Roosevelt met with Saudi King Abdul Aziz aboard a Navy destroyer in the Suez Canal to cement the fledgling US–Saudi relationship. It continues to this day. Any attempt to disrupt energy supplies is a direct threat to the lifeblood of the industrial democracies. In the Cold War context, it meant containing Soviet expansionism—not just the Soviet threat to those along its borders, but the larger Soviet threat to the alliances. Protecting energy supplies was a shared goal across the industrial democracies even though there were differences on how to make good on that goal. In the simplest terms, the industrial democracies were motivated by the basic idea of making friends and keeping threats at a distance.

Somewhat ironically, none of the treaties were called upon to help with the defense of the United States during the Cold War. This is not to say they were not valuable. Enormous effort went into maintaining the alliances and enormous funds were expended to support them. Perhaps the best measure of the Cold War alliances is the fact that the United States never fought a war with the Soviet Union, a remarkable historical achievement given the many times, which we will see, when a war between the two Great Powers might have been fought. Article 5 of the NATO treaty was first invoked on September 12, 2001, in the aftermath of the 9/11 attacks on the United States. To the extent that US strategy aimed to keep threats at a distance, the system of alliances and relationships worked remarkably well.

In time, as the American and Soviet nuclear stockpiles grew, attention turned to managing the competition in ways that were more stabilizing rather than less so. Particular attention focused on finding ways to dissuade either party from considering a surprise disabling attack—known at the time as a "bolt from the blue"—and to bring more predictability to the larger arms race. This led to a number of treaties controlling the growth in weapons stockpiles, eliminating certain classes of weapons, limiting defenses against ballistic missiles, banning tests of nuclear weapons, and, importantly, limiting the proliferation of nuclear weapons and related technologies. This prompted an alphabet soup of abbreviations like SALT, START, ABM, INF, CTBT, and NPT.[19] Why all this matters is wrapped up in the arcane logic of the Cold War. The enormous destructive power of nuclear weapons meant each side feared a disarming attack—an attack from the other party that destroyed its ability to retaliate and potentially destroyed its industrial capacity and ability to govern. The idea that a hundred or two hundred weapons would suffice to deter the other party only worked if both parties had great confidence the hundred or two hundred weapons, along with the political and military leadership needed to orchestrate the response, would survive an attack. Neither party had such confidence, so both parties continued to build additional weapons. Although it was sometimes said that the additional weapons would only be needed to "bounce the rubble"—that is, to reattack targets that already had been attacked—there was an underlying rationale for how the additional weapons would be used. Arms control was the antidote to the arms race.

The focus on arms control gave rise to a whole new group of experts, many of them civilian, who would generate ideas, participate in meetings and conferences, and staff the painstaking efforts to arrive at arms control agreements. From this cadre would arise a host of new characters and personalities, including Henry Kissinger, who was a Harvard professor active in the early arms control dialogue before he went on to his more famous role as national security adviser and secretary of state

[19] SALT (Strategic Arms Limitation Talks), START (Strategic Arms Reduction Treaty), ABM (Anti-Ballistic Missile) Treaty, INF (Intermediate-Range Nuclear Forces) Treaty, CTBT (Comprehensive Test Ban Treaty), and NPT (Non-Proliferation Treaty).

to Richard Nixon and Gerald Ford and adviser to all presidents since. Others, like Zbigniew Brzezinski, national security adviser to Jimmy Carter, and Brent Scowcroft, national security adviser to George H. W. Bush, were products of the same system. Closer to the present, Condoleezza Rice and Steven Hadley, both national security advisers to George W. Bush, had backgrounds in the larger arms control community. Ashton Carter, defense secretary to Barack Obama, was a product of the arms control community as well.

New centers emerged at major universities like Harvard, Yale, Princeton, Columbia, MIT, Stanford, and Johns Hopkins. Education in new specialties was required, and the university programs responded to the larger market signal. As the machine grew larger and more complicated, the university system was ready to train the talent the machine demanded. This all became part of the larger national security system.

Various lessons piled up over time, from a series of both failures and successes. Each, in a way, led to a retooling of the warning and action machines. Not so much an overhaul, but a string of important changes that had lasting effects, these successes and failures fill the pages of Cold War analyses and now Cold War histories. Our focus on reviewing these successes and failures is to illustrate the impact they had on the warning and action machines and how they, cumulatively, have brought us to the current moment.

North Korea's June 1950 invasion of South Korea marked the first real test of the modern warning and action machines. Both were found wanting, and deeply so. North Korea's attack came after five years of simmering tensions, and American forces were unprepared for combat just a few short years after the major victories of World War II.[20] Task Force Smith was deployed to help slow the North Korean advance and was quickly overrun. US forces were outgunned and outmaneuvered and soon fell into retreat. The amphibious landing at Inchon that came several months later marked a stark turnaround in the war. But for generations to come, Task Force Smith would be the rallying cry of those advocating for trained and ready forces. To this very day among Army and Marine Corps advocates, Task Force Smith is remembered with a sense of "never again." Future secretaries of defense, including Dick Cheney and Jim Mattis, would refer to Task Force Smith when arguing the case to build and sustain military readiness.

The experience gained from the Korean War meant the United States would no longer rely on mobilizing its military forces at the time of war, which was the historical tradition, but instead would rely on large, standing forces, forces capable of being deployed on a moment's notice, sometimes within hours of being alerted to action. Training had to be continuous because if forces were to be ready, they had to be trained. Not every part of the military had to be trained to the same level of readiness at the same time, but some portion of the military services had to be trained and ready all the time. During the Cold War, in that era when the Soviet Union could launch at a moment's notice, nuclear forces were kept at the highest level of readiness and alert. Long-range bombers had to be capable of loading weapons and flying with little or no notice. This was known as strip alert. Readiness meant that nuclear missile submarines were at

---

[20] "Korean War," Dwight D. Eisenhower Library, Museum, and Boyhood Home, https://www.eisenhowerlibrary.gov/research/online-documents/korean-war.

sea and on station every day of every month of every year, and land-based missile forces had crews on station around the clock should the president give the order to launch. No other part of the fighting force maintained this level of readiness, but every part was expected to be ready to deploy within hours if not days of receiving an order. This meant that every part of the system had to have a backup, and the most important parts of the system had to have backups to the backups. All this required people, know-how, and money. As a result, whereas defense spending had fallen sharply as World War II came to a close, it rose again sharply when the country decided to maintain a standing, ready force. In many ways, this is the legacy of Task Force Smith—what it meant to not be ready in a time of need.

The Bay of Pigs in Cuba marked another important milestone in the maturing of the warning and action machines. This incident revealed the vulnerability that can arise in passing the baton from one presidential administration to another. When John Kennedy was elected president, the CIA, under the direction of Allen Dulles, brother of former secretary of state John Foster Dulles, had developed plans to topple Fidel Castro, the young Cuban firebrand—then only thirty-four years old—by relying on a mix of Cuban refugees supported by US forces. Dulles presented his plan to the brand-new Kennedy administration, and it was approved within days of Kennedy's inauguration. The subsequent operation proved a colossal failure and incredible embarrassment for the new president. This despite Eisenhower's caution that "the plan could not be hatched until a government-in-waiting was in place."[21] Eisenhower assumed

---

[21] M. Kent Bolton, *The Rise of the American Security State: The National Security Act of 1947 and the Militarization of US Foreign Policy* (Santa Barbara, CA: ABC-CLIO, 2017), 68–72.

the new administration would understand the risks before it set the CIA's plans in motion. The resulting disaster set the stage for US–Cuban relations to the present and is cited by many as a proximate factor in the Cuban Missile Crisis, which followed in just under two years. It also represents perhaps the first big step in drawing decision-making closer and closer to the White House. No president wants to be presented with a foreign policy disaster during the first year in office, particularly a disaster of his own making. The lesson Kennedy took from the Bay of Pigs was that he not only wanted his stamp on all key decisions, he wanted close White House oversight of all the implementing actions. Kennedy quickly replaced Dulles as CIA director, wanting "his people" in charge of the national security machine. This would become a recurring theme for nearly all subsequent presidential administrations.

The Cuban Missile Crisis served as the single biggest test of the warning and action machines during the Cold War. Despite huge expenditures on surveillance and warning systems, including a new fledgling space program, the Kennedy administration was surprised that Soviet leader Nikita Khrushchev was installing missiles in Cuba capable of delivering nuclear warheads to the United States from just a short distance away. John McCone, Kennedy's new CIA director, was alone among Kennedy's advisers in warning that Khrushchev was planning such a bold step. In one of the oddities of history, McCone was on travel for his honeymoon during the crucial time when the mounting intelligence needed an advocate in the Kennedy National Security Council. Others didn't see it, or didn't believe it, and thus the threat continued to grow.

As later released in CIA records, "McCone's hunch turned into reality when U-2 spy plane photographs revealed construction of Soviet medium-range ballistic missile sites near San Cristobal, in northern Cuba."[22] The subsequent thirteen days became a turning point in the Cold War, unquestionably the closest that either the United States or the Soviet Union ever came to nuclear war. Kennedy and Khrushchev engaged in communications throughout the crisis. At one point, Khrushchev sent the following message to Kennedy: "If there is no intention to doom the world to the catastrophe of thermonuclear war, then let us not only relax the forces pulling on the ends of the rope, let us take measures to untie that knot. We are ready for this."[23] The crisis was resolved when Kennedy committed not to invade Cuba, Khrushchev removed the missiles from Cuba, and Kennedy later removed US missiles and warheads from Turkey. The warning and action machines had survived the greatest test to date. A direct communications link would be established between the White House and the Kremlin known as the "hotline."

Even though Kennedy acted on warning and the crisis was ultimately resolved, Kennedy did not know critical information at the time. As Graham Allison, author of the most famous book on the Cuban Missile Crisis, *Essence of Decision*, reminds us, the United States and Russia are once again going toe-to-toe over the war in Ukraine:

> How could the dynamics in 1962 have led to nuclear war? Analysts of this crisis have identified more than a

---

[22] Excerpts from this *Washington Post* article on CIA records (linked) reveal the chaotic behind the scenes in Kennedy's NSC during the Cuban Missile Crisis. https://www.washingtonpost.com/wp-srv/inatl/longterm/cuba/stories/history101992.htm.

[23] "The Cuban Missile Crisis: October 1962," Office of the Historian, https://history.state.gov/milestones/1961-1968/Cuban-missile-crisis.

> dozen plausible paths that could have led to the incineration of American cities. One of the fastest begins with a fact that was not even known to Kennedy at the time. The core issue for Kennedy and his associates was preventing the Soviets from installing operational medium- and intermediate-range nuclear missiles in Cuba that could strike the continental United States. They were unaware, however, that the Soviets had already positioned more than 100 tactical nuclear weapons on the island. Moreover, the 40,000 Soviet troops deployed there had both the technical capability and the authorization to use those weapons if they were attacked.[24]

So, yes, the resolution came to be viewed as a triumph, but the incident was actually a much closer call than anyone imagined, even those who knew what a close call it was. Even well-tuned crisis and action machines won't surface all the relevant information needed for making important decisions. The idea that Kennedy and Khrushchev stood toe-to-toe and Khrushchev blinked has become part of American foreign policy lore. A more appropriate conclusion would be that two statesmen backed the world away from calamity. The machine did its job. The leaders in charge did a better job.

There is another lesson from the Cuban Missile Crisis that gets much less attention. The crisis had a beginning, a middle, and an end. Problem identified, action taken, problem resolved. In a way, it left a blueprint for successful crisis management. Volumes have been written about it over the decades. Robert Kennedy, the president's brother and attorney general, wrote his

[24] Graham Allison, "Putin's Doomsday Threat," *Foreign Affairs*, April 2022.

account in a book titled *Thirteen Days*. But few national security crises have such a straightforward narrative. Most problems go on for months, years, and even decades. They end up being managed and not resolved. And here's the problem when crises are managed, not resolved, in the White House: With a crisis that goes on for days, the president and his team can devote full attention to it. But a crisis that goes on for weeks, months, or years cannot possibly command the president's full attention. Plus, in time other problems and crises arise. No White House is built for multiple crises or for multiples of multiples. Still, the Cuban Missile Crisis blueprint remains the prevalent action model—pull the issue and the decision makers into the White House and sustain its attention until it is resolved. In a world where problems come in multiples and few issues actually get resolved, it is not uncommon for decision makers across the warning and action machines to spend much of their daylight hours in the White House and their early mornings and late evenings back at their home offices. A different model for managing crises has yet to emerge.

The war in Vietnam served as the next great test of the national security apparatus and would have a lasting impact. Whereas the stakes for America were never as high as in the Cuban Missile Crisis, the human toll was far, far greater in Vietnam. More than fifty-eight thousand Americans were killed in Vietnam, and an estimated two million North and South Vietnamese were killed. Vietnam was the first US war broadcast daily into American living rooms. It was also the first US war where key decisions on the day-to-day conduct of the war came directly from the White House. President Lyndon Johnson, surrounded by advisers, famously examining maps of troop movements and potential bombing targets was legendary.

As an Army lieutenant colonel, H. R. McMaster, who would later serve as Donald Trump's national security adviser, wrote a PhD dissertation on what he termed "dereliction of duty," a broad indictment of the US military leadership's unwillingness to confront Johnson and his advisers over the conduct of the war. "Dereliction of Duty" became required reading at the military war colleges, and that's true to this day.

It was during the Vietnam era that Defense Secretary Robert McNamara brought the tools of industry to the Pentagon. He introduced more structured processes for planning and budgeting, a more business-like approach to running the world's largest bureaucracy. This, in turn, produced the modern planning, programming, and budgeting system that consumes endless amounts of staff time at the Pentagon to this day. It was McNamara who also insisted on metrics for measuring progress in military preparations and actual warfighting. He hired a team of systems analysts to introduce the latest management tools. Modern approaches to cost-benefit analysis were introduced during the McNamara era and continue to be used today. They are an upside to his approach. The use of body counts as a measure of warfighting effectiveness was also introduced. This was a distinct and highly misleading downside.[25]

The US military had to be remade to fight in Vietnam. The forces required to halt an invading army or drive an army from a contested area are very different from those needed to help a partner, in this case, an especially weak partner, fight off an insurgent force that is being supported by a large standing

---

[25] For a valuable description of the early effort to introduce a more coherent planning and budgeting system to the Pentagon, see Alain C. Enthoven and K. Wayne Smith, *How Much Is Enough?* (Santa Monica, CA: RAND Corporation, 1971).

army—the Viet Cong and the North Vietnamese Army. The US Army and Marines, in particular, created new counterinsurgency doctrine to guide their efforts, drawing heavily from lessons from the British and French when they tried to maintain toeholds in their former colonial territories. New approaches to fighting were tested; new kinds of equipment were developed that involved US military forces operating within and living alongside the civilian population of South Vietnam. The war was a mixture of both pacifying areas and large, pitched battles in now famous areas like Khe Sanh, Ia Drang, and Hue.

The anguishing experience in Vietnam produced a whole new literature, fiction and nonfiction alike. In a fascinating analysis little known outside academic circles, Leslie Gelb and Richard Betts wrote a volume titled *The Irony of Vietnam: The System Worked*, their own postmortem on the warning and action machines. Gelb and Betts argue that the machinery had operated as intended; that the consensus that formed the basis for post–World War II foreign policy—the containment of communism—had held; that differences of opinion, among elites and among the public, were accommodated by compromise, thus taking the options of steep military escalation or disengagement off the table; and, importantly, that decisions had been made without illusion regarding the actual odds for success. They conclude that Presidents Kennedy and Johnson knew the United States was not in a position to win the war in Vietnam. Each commander in chief focused his attention instead on not losing.[26] The authors quote Kennedy's adviser Theodore Sorensen, who recalled that in 1963 Kennedy "was simply going to weather it out, a nasty,

---

[26] Leslie Gelb and Richard Betts, *The Irony of Vietnam: The System Worked* (Washington, DC: Brookings Institution Press, 1979).

untidy mess to which there was no other acceptable solution."[27] Interestingly, Gelb and Betts begin their analysis with something of an apology: "The title of this book must strike any intelligent reader, at first glance, as ridiculous."[28]

Vietnam was a war, to be sure, but it was also more than that. The commitment to the war entailed supporting a weak government that was under attack by an organized, committed foe. US military forces were helping to fight off the attack from the North, but other parts of the warning and action machines needed to be mobilized to support the failing government of South Vietnam. This involved providing economic and development assistance that had to fall in on the crumbling colonial structure the French had left behind. Warning and action machines veteran Robert Komer wrote about this in a volume with the memorable title *Bureaucracy Does Its Thing.*[29] Komer talks about how every part of the machine did its job, but the machine as a whole was not up to the task. He points out how South Vietnam was a weak partner that played a weak hand amazingly well. He notes that one thing the United States was not willing to do was take over governing South Vietnam, which is what might have been necessary to actually succeed, but it smacked of colonialism and was outside the bounds of what US decision makers were willing to consider. So, for more than a decade, the United States continued to play a losing hand at the cost of American lives and treasure. The United States was unwilling to take control of the territory it was defending, and the ally it was defending was incapable of

[27] Gelb and Betts, *Irony of Vietnam*, 196.

[28] Gelb and Betts, *Irony of Vietnam*, 1.

[29] Robert Komer, *Bureaucracy Does Its Thing: Institutional Constraints on U.S.-GVN Performance in Vietnam* (Santa Monica, CA: RAND Corporation, August 1972).

maintaining control. The lesson from this experience was "no more Vietnams," which was etched rather deeply into the American psyche. This sentiment was captured in the "Nixon doctrine" that followed from the Vietnam experience. The new policy, at its essence, maintained that the United States would help its friends and partners across the globe by supplying equipment and assistance, but it would not do their fighting for them.

There is another thought to consider regarding Vietnam. In contrast to the seemingly endless compounding of threats we see today, Gelb and Betts offer a fascinating observation on Lyndon Johnson's choices in Vietnam:

> Three facts conspired to make it easier for Johnson to take the plunge on the assumed importance of Vietnam.... First, the world was a safer place to live in and Vietnam was the only continuing crisis. Europe was secure. NATO troubles were relatively minor. The Sino-Soviet split had deepened. Mutual nuclear deterrence existed between the superpowers. Second, the situation in Vietnam was more desperate than it had ever been. If the United States had not intervened in 1965, South Vietnam would have been conquered by the Communists. Third, after years of effort the U.S. conventional military forces were big enough and prepared enough to intervene. Unlike his predecessors, Johnson had the ready military capability to back up his words.[30]

The force that hadn't been ready when South Korea came under attack was now ready for a different type of mission.

---

[30] Gelb and Betts, *Irony of Vietnam*, 196.

Perhaps only George W. Bush would face a similar decision calculus prior to the 2003 invasion of Iraq—the world appearing to be safer and the choices his to take. But Bush would find himself enmeshed in the middle of a civil war far from home with few good choices except to escalate. Like Johnson, he believed he had to go big in order to get out.

The Vietnam experience was formative for other reasons, too. It is where future leaders like Colin Powell, John Kerry, and Chuck Hagel formed their views on the role of American power in the world.[31] It is where other future leaders like Bill Clinton, Dick Cheney, and Donald Trump elected not to serve in the military and would later be judged on that choice when they decided to enter public life. Add George W. Bush, who served in the Texas National Guard but whose service was seen as a way to avoid being sent to Vietnam.

By the late 1980s, the Vietnam experience seemed distant and the Reagan defense buildup was bearing fruit, though some elements, like the Star Wars expenditures, remained controversial. After a series of leadership successions, the Soviet Union had a new, youthful leader named Mikhail Gorbachev, and perestroika was in the air. Although the United States was finding itself more and more engulfed in the Middle East, there was much less concern about the Soviet threat. The Soviet Union had its own problems in Afghanistan in a fight against insurgents who were being supported by the United States. A sense of hope was emerging as President Ronald Reagan challenged Gorbachev to "tear down this wall" that was separating

[31] All future cabinet members: Powell and Kerry as secretary of state; Hagel as secretary of defense. Other recent cabinet members like Robert Gates and Leon Panetta served in the military but were not deployed to Vietnam. Panetta also served as CIA director.

East and West Berlin. Even the famous REFORGER exercise, which was designed during the Vietnam era to demonstrate the ability of US forces to reinforce the NATO armies in a crisis, became something of a festival in Germany, where US troops could indulge themselves in German beer and soft pretzels. REFORGER was often held in September, which coincidentally overlapped with Munich's Oktoberfest.

In this period, one particular incident stands out for the attention it brought to the action machine. In response to the taking of American hostages in Iran, President Jimmy Carter directed a rescue operation named EAGLE CLAW to free the hostages and bring them back to the United States. Planning began in November 1979 and the operation was carried out in April 1980. The result was a calamity in the desert outside Tehran. Eight Americans and one Iranian were killed. The resulting investigation revealed poor planning and execution across the military services. EAGLE CLAW served as motivation for the Goldwater–Nichols reforms that followed several years later that altered the chain of command so that orders flowed from the president to the secretary of defense to the combatant commander in charge of the operation.[32] No other part of the action machine, including the military services, stood between the president and his chief civilian adviser and the military commander in the field. The reforms also led to the establishment

[32] Often hailed as one of the hallmarks of the Goldwater–Nichols reforms, the chain of command flows from the president to the secretary of defense to the combatant commanders. The military services are responsible for organizing, training, and equipping military forces, but they are not responsible for conducting military operations. That responsibility lies with the combatant commanders—US Northern Command, US Southern Command, US Africa Command, US European Command, US Central Command, US Indo-Pacific Command, US Special Operations Command, US Strategic Command, US Space Command, US Cyber Command, and US Transportation Command.

of US Special Operations Command, or SOCOM, which would play such a prominent role after the 9/11 attacks.

None of this is to suggest that the Cold War wasn't fraught until the end. The Soviet Union still possessed a vast array of nuclear weapons, and although it engineered a soft landing by tossing in its cards and allowing the former Warsaw Pact members to strike out on their own, there was always the possibility of a hard landing, such as the collapse of one of the former Soviet regimes or loss of control of the many nuclear weapons that were stationed across the former Soviet territory. In one of perhaps the least heralded successes of the twentieth century, the warning and action machines worked together to bring the Cold War to a quiet and successful end. That required an astounding feat of statesmanship, a virtuoso effort if there ever was one. The amount of coordination and collaboration that had to occur among allies and partners, to say nothing of former adversaries, goes almost beyond description. Germany was united and new countries were born. Hope and excitement abounded, leading one analyst to wonder whether the world was witnessing the "end of history."[33]

STILL ON PANORAMA, BUT SEARCHING FOR A NEW LENS

Almost immediately, a new debate arose about the purposes of American power. Some argued for a return to a more traditional foreign policy—with fewer entanglements and a smaller military. The United States had mobilized to confront a global Soviet threat, but it was time to bring that talent and energy back to solving problems closer to home. Others argued it was

---

[33] Francis Fukuyama, *The End of History and the Last Man* (New York: Free Press, 1992).

exactly the time for the United States to secure the gains that had accrued during the Cold War and ensure that big, new problems did not arise. In a famous leaked document of the era, Dick Cheney and his policy team at the Pentagon were sketching out a new strategy that called for "preclud[ing] hostile, nondemocratic domination of a region critical to our interests," or as the *New York Times* reported, "insuring no rivals."[34] This debate would continue through the 1990s as the United States confronted Iraq in the Middle East, deployed troops to forestall genocide amid warring factions in Somalia, intervened on multiple occasions in the Balkans, and even invited new members into the NATO alliance.

In some ways, the First Gulf War, which followed the August 1990 Iraqi invasion of Kuwait, showed something of the culmination of the warning and action machines. Iraq's invasion of Kuwait was more than one regional power toppling the government of another. It represented the consolidation of control over a significant amount of the world's energy supply. Left unchallenged, Iraq could go on to threaten Saudi Arabia, which sat atop the largest energy reserves in the world.

If the end of the Cold War was a demonstration of soft power—which brought a hard-fought competition to a quiet end—the First Gulf War was a demonstration of America's newly tooled hard power, the modern remaking of the action machine. Soon after the invasion, there was a debate in the United States about whether the invasion may have been deterred had the warning machine signaled Saddam Hussein's

[34] Patrick E. Tyler, "U.S. Strategy Plan Calls for Insuring No Rivals Develop," *New York Times*, March 8, 1992, https://www.nytimes.com/1992/03/08/world/us-strategy-plan-calls-for-insuring-no-rivals-develop.html. A. Hoehn was part of the drafting team.

intent earlier. But intentions are hard to detect. What the warning machine did provide was detection of Iraqi troop movements and signs of a pending attack. On July 26, 1990, Charlie Allen, then national intelligence officer for warning, sounded the alarm that Saddam Hussein's forces were poised to attack. Little happened over the next week until Hussein gave the go-ahead and Iraqi forces quickly overran Kuwait.[35] It is difficult to know whether acting earlier would have deterred Hussein. What is known is that the warning machine sounded the alarm, and the action machine did not swing into action until after Kuwait was under Iraqi control.

When activated, the action machine then stepped up to full gear. George H. W. Bush was president. Brent Scowcroft was national security adviser. James Baker was secretary of state. Dick Cheney was secretary of defense. And Colin Powell was chairman of the Joint Chiefs of Staff. All were seasoned veterans of the warning and action machines. Bush himself was a former CIA director. When Bush decided Iraq's invasion "will not stand," he put the larger machine into motion. Baker mounted a global hop-scotching effort to build an extraordinary coalition to push the invading forces out of Kuwait. The secretary of state amassed not only diplomatic and military support but also financial support to offset the costs. Some called this "operation tin cup" because of the vast financial commitment that Baker was able to amass in support of the war.[36]

---

[35] As described in Richard A. Clarke and R. P. Eddy, *Warnings: Finding Cassandras to Stop Catastrophes* (New York: HarperCollins, 2017).

[36] As of February 11, 1991, the United States had amassed $50 billion in pledges to pay wartime costs. David E. Rosenbaum, "War in the Gulf: Financing; U.S. Has Received $50 Billion in Pledges for War," *New York Times*, February 11, 1991, https://www.nytimes.com/1991/02/11/world/war-in-the-gulf-financing-us-has-received-50-billion-in-pledges-for-war.html.

Cheney and Powell took charge of the military response. Once they secured agreement from Saudi Arabia to allow forces to deploy to the kingdom, American fighter jets and paratroopers soon followed, establishing an initial deterrent force and allowing for substantial reinforcements to follow. "Desert Shield" was soon in place, and "Desert Storm" followed a few months later. Iraqi forces were ejected from Kuwait after thirty days of punishing aerial bombardment followed by a ground invasion featuring the "left hook" so famously depicted by General Colin Powell in the Pentagon press room. The F-117 stealth bomber made its debut, as did a recently modernized Army featuring new tanks, helicopters, and troop carriers.

Looking back, it is no surprise the most modern military of the time would hold sway over a much lesser power. But the shadow of Vietnam cast doubt over even the most knowledgeable leaders of the time. Senator Sam Nunn, then chairman of the Senate Armed Services Committee, voted against the war, partly out of concern that the cost in American casualties would not be worth the effort. Even some responsible for the actual fighting harbored lingering doubt about whether the retooled machine, which started up after Vietnam and achieved full motion during the Reagan defense buildup, would actually work. In a memorable quote from one of the pilots to first fly the F-117 into Iraqi airspace, he confessed, "Well, I sure hope to God that stealth shit really works."[37] Fortunately for him and those flying with him, it did.

---

[37] Ben Rich and Leo Janos, *Skunk Works: A Personal Memoir of My Years at Lockheed* (Boston: Back Bay Books, 1996), 100.

More than twenty-five years after the First Gulf War, Colin Powell was asked for his reflections. They are worth considering at some length:

> It was the only time in my career or in, frankly, most of American military history, where a chairman can say to the president of the United States, I guarantee the outcome. And the reason I could guarantee that outcome is that the president gave us everything we asked for. In a relatively short period of time, the Iraqi army was no longer in Kuwait, and the government had been restored.
>
> But the best part from my perspective is the way in which the American people saw this operation. And they had been told that tens of thousands might be killed. They were worried about this volunteer army that had never been in this level of combat before. And they were absolutely joyful at the results. And they threw parades for our troops. And it just refreshed my memory that a classic military theory says, make sure you know what you're getting into.[38]

Colin Powell was describing the national security machine working at its best. The war in Iraq was almost a perfect bookend to the Cold War. If the Cuban Missile Crisis showed the warning and action machines working in tandem to avoid

---

[38] Tony Lombardo, "Q-and-A: Colin Powell on Vietnam Service, Iraq and Afghanistan, and Black History Month," *Military Times*, January 31, 2017, https://www.militarytimes.com/military-honor/black-military-history/2017/02/01/q-and-a-colin-powell-on-vietnam-service-iraq-and-afghanistan-and-black-history-month/.

nuclear war, so, too, did the First Iraq War show the warning and action machines working in tandem to contend with a new type of regional threat.

Almost.

Although Iraqi forces were evicted from Kuwait and the Kuwaiti government was returned to power, which remains true to this day, US forces stayed in the Middle East, first to monitor the truce that was struck and then to enforce punishing sanctions and military restrictions, including no-fly zones, on the Iraqi government. The continuous presence of US forces in the Arabian Gulf would be grating to the local hosts and would set the stage for the second war with Iraq that occurred twelve years later.

A large number of modern-day figures emerged from the First Gulf War. Dick Cheney and Colin Powell were secretary of defense and chairman of the Joint Chiefs of Staff, respectively. Paul Wolfowitz served as undersecretary of defense for policy. Richard Armitage was special envoy to King Hussein of Jordan during the war. Steve Hadley was a deputy to Wolfowitz at the Pentagon. Cheney, of course, would become vice president to George Bush nine years later. Powell would become secretary of state. Armitage would be deputy to Colin Powell at the State Department. Wolfowitz would be deputy to Donald Rumsfeld at the Pentagon. Hadley would be deputy to Condoleezza Rice on the National Security Council staff. He would later become national security adviser.

Others who emerged later were also shaped by the First Gulf War experience. Mark Esper, who would serve as Donald Trump's second secretary of defense, was an infantry officer with the Army's 101st Airborne Division in the First Gulf War. President Biden's selection for defense secretary, Lloyd Austin, was a young Army officer during the Gulf War. Austin later rose to the rank of general and served two tours in Iraq during the Second Gulf War, one as a commander of the Army's 3rd Infantry Division during the 2003 invasion of Iraq.

For all these leaders, if the First Gulf War showed the strength of the warning and action machines, the shocking September 11, 2001, attack on New York and Washington showed their weakness.

EVEN WHEN WARNING WAS ACCURATE, ACTION DID NOT FOLLOW UNTIL TOO LATE FOR TOO MANY

When Russian forces rolled into Ukraine to launch an unprovoked invasion ordered by Vladimir Putin, the world witnessed a refugee exodus not seen since World War II—and atrocities, likely war crimes, also not seen since Hitler's time. Tragically, the invasion of Ukraine in 2022 merely displaced another conflict in the former Yugoslavia—the wars of the 1990s—as the greatest human tragedy brought by armed conflict in Europe since World War II.

Serbia's war of aggression in Bosnia was a medieval-style conflict fought with twentieth-century weapons and was marked by siege, razing of villages, mass murder, and mass rape.

And the American intelligence community saw it coming. In October 1990, the director of central intelligence received a National Intelligence Estimate entitled "Yugoslavia Transformed," which predicted with grim clarity the likelihood that Yugoslavia, once a heterogenous success story, a nonaligned communist country with one of the highest standards of living in Eastern Europe, would collapse into ethnic and religious violence: Serbian Eastern Orthodox versus Croatian Catholic versus Bosnian Muslim.

The report, stamped SECRET, was declassified in 2006, so we can review its findings with the benefit of heartbreaking hindsight.[39]

To be sure, the National Intelligence Estimate contained all the caveats that would give leaders options to not act. "Yugoslavia will cease to function as a federal state within one year, and will probably dissolve within two," the report states in its executive summary—a timeline that proved correct. It predicted that most violence would be centered in Serbia's Muslim region of Kosovo—but that timeline was off, as the Kosovo War did not begin until 1998.

"Serious intercommunal conflict will accompany the breakup and will continue afterward," the report said. Correct. "A full-scale, interrepublic war is unlikely." That was falsely optimistic.

But if a reader merely continues to page 9, the National Intelligence Estimate contains a prediction that was prophetic: "The most plausible scenario for interrepublic violence is one in which Serbia, assisted by disaffected Serbian minorities in the other republics, moves to reincorporate disputed territory into a greater Serbia, with attendant and bloody shifts in population."

That is exactly what happened. What didn't happen was any muscular American actions to save the lives of innocent Muslim men being herded into death camps or Muslim women into mass-rape camps. The war was a slow-burning fuse; options to extinguish it were complicated and messy. Even with that dire prediction, official Washington failed to articulate, or even see, a vital national security risk that required

---

[39] "Yugoslavia Transformed" (National Intelligence Estimate, Director of Central Intelligence, October 18, 1990), https://www.cia.gov/readingroom/docs/DOC_0000254259.pdf.

intervention. There was little appetite for interventions in the Balkans, historically home to ethnic and religious blood feuds dating back to hurts that were centuries old. Locals joked, though, that only the odd-numbered world wars began in Sarajevo, the Bosnian capital.

Only when lightly armed Dutch peacekeepers operating under a United Nations mandate were seized and humiliated by Serbian forces en route to massacre more than eight thousand Muslim men at Srebrenica in July 1995 was the Clinton administration shamed into finally leading Europe to act.

A coordinated Western bombing campaign brought the combatants to an Air Force base in Dayton, Ohio, for peace talks to end the conflict, or at least the overt violence. By then, the death toll had reached one hundred thousand with about two million people displaced. Despite clear warnings from within the US government, Washington and Western capitals did little to prevent the greatest loss of life in European conflict since World War II.

It is often said that September 11, 2001, began like any other fall day in Washington and New York—blue skies, pleasant fall temperatures, people streaming to work in shirtsleeves, kids back at school.[40] Just the day before, Don Rumsfeld had lambasted the Pentagon bureaucracy as the "new enemy" that was holding back his efforts to transform the US military. He made clear his focus was on the systems—the machine—and

---

[40] Hoehn was present in the Pentagon on 9/11, as a member of Rumsfeld's policy staff. Shanker, a national security reporter for the *New York Times*, was on public transportation en route to his desk at the Pentagon when the airliner struck the building.

not the people, but truth be told, part of his ire was with the people:

> The topic today is an adversary that poses a threat, a serious threat, to the security of the United States of America. This adversary is one of the world's last bastions of central planning. It governs by dictating five-year plans. From a single capital, it attempts to impose its demands across time zones, continents, oceans and beyond. With brutal consistency, it stifles free thought and crushes new ideas. It disrupts the defense of the United States and places the lives of men and women in uniform at risk.
>
> Perhaps this adversary sounds like the former Soviet Union, but that enemy is gone: our foes are more subtle and implacable today. You may think I'm describing one of the last decrepit dictators of the world. But their day, too, is almost past, and they cannot match the strength and size of this adversary.
>
> The adversary's closer to home. It's the Pentagon bureaucracy. Not the people, but the processes. Not the civilians, but the systems. Not the men and women in uniform, but the uniformity of thought and action that we too often impose on them.[41]

Rumsfeld spent much of his first year frustrated that he couldn't impose his will on the larger machine called the

[41] Donald Rumsfeld, "Remarks as Delivered by Secretary of Defense Donald H. Rumsfeld, the Pentagon, Monday, September 10, 2001," A Government of the People, https://agovernmentofthepeople.com/2001/09/10/donald-rumsfeld-speech-about-bureaucratic-waste/.

Pentagon and its many appendages. It was no longer the department he remembered leading when he was in the job twenty-five years before. Every layer of management had additional layers under it. He felt himself alone, at first, trying to direct the largest bureaucracy in the world with few direct subordinates—the many undersecretaries, assistant secretaries, and deputy assistant secretaries who help the secretary of defense run the massive bureaucracy. This was true, as well, of the military departments. He had to wait longer than four months for the secretaries of the Army, Air Force, and Navy to be confirmed for their jobs. The machine was now much larger than when he left his post twenty-five years earlier. The Pentagon, which was once filled with phone operators, typists, and clerks, was now filled with staff officers working for the secretary of defense, the chairman of the Joint Chiefs of Staff, and all the military services.

This was less true for the intelligence community. George Tenet had been appointed director of central intelligence in 1997 by President Bill Clinton and was kept in his role by George W. Bush. The intelligence community had far fewer presidential appointees than the Defense Department, or any other cabinet department, for that matter. Osama bin Laden and al-Qaeda were very much in Tenet's sights after the failed World Trade Center attack in 1993, various attacks against US military personnel in Saudi Arabia, and, while Tenet was CIA director, the 1998 attacks against US embassies in Kenya and Tanzania. Soon after the embassy attacks, Clinton directed cruise missile attacks against al-Qaeda facilities in Sudan.

What Tenet didn't know was that prior to 9/11 al-Qaeda cells were operating within the United States. Although the warning machine had been built and tuned over time to search for threats abroad, it had no counterpart for monitoring threats within the United States. The FBI serves as the federal law enforcement arm, but the FBI was not organized to gather and provide intelligence from within the United States. And should the FBI identify potential threats within the United States, it had no authority or obligation to share that information with the vast machinery responsible for foreign intelligence that was under Tenet's direction. Even if the terrorist cells had been detected, law enforcement would need evidence of al-Qaeda's plans before it could act. The FBI was held to a different standard than the CIA.

This was the vast hole in the warning and action machines that al-Qaeda exploited in the 9/11 attacks.

In the aftermath of 9/11, all attention focused on patching the holes in the warning and action machines. Within days of the attack, President Bush appointed former Pennsylvania governor Tom Ridge to a new White House position as director of the Office of Homeland Security. By June 2002, Bush proposed the creation of a new cabinet department, the Department of Homeland Security, and by November 2002 Bush's proposal was passed into law. The new agency knit together twenty-two federal departments and agencies. Tom Ridge became the first secretary of the new department.

From the beginning, the new Department of Homeland Security, or DHS, struggled to knit together the many disparate functions that had been brought under a single authority. With responsibilities ranging from customs and border control to cyber security and domestic disaster assistance, the new department struggled to find an identity and department-wide culture. That continues to this day. Not all agreed at the time with the recommendation to pull the disparate functions together, though many observers noted the agencies assembled in the new department were orphans in their previous homes. In a critique written years later, Richard Clarke, longtime insider and counterterrorism coordinator for Clinton and Bush, argued that DHS should be split into two organizations, one focused on safety and protection, the other focused on customs, immigration, and borders.[42] Even twenty years later, the roles of the department are still being sorted out. DHS has yet to go through the type of reorganizations the Defense Department went through in its earlier history.

The intelligence community was quickly retooled to focus on terrorist threats. A Director of National Intelligence was established to be the coordinator of the nation's many intelligence activities. This role was formerly undertaken by the Director of Central Intelligence.

The National Counterterrorism Center (NCTC)—originally organized as the Terrorist Threat Integration Center (TTIC)—was established over the objections of various parts of the intelligence community that were concerned they would need to surrender people and resources that were important for their original duties. Here's how the NCTC describes itself: "Before the establishment of TTIC, individual federal departments and agencies (largely CIA and FBI) provided the President their own assessments of the terrorist threat. In effect, the White House was being forced to synthesize Community reporting

---

[42] Richard A. Clarke, "Opinion: Dismantle the Department of Homeland Security," *Washington Post*, July 30, 2020, https://www.washingtonpost.com/opinions/dismantle-the-department-of-homeland-security/2020/07/30/24ef8ba0-d279-11ea-8c55-61e7fa5e82ab_story.html.

and draw its own conclusions. This was among the first systemic issues that TTIC would be tasked to address and would be critical, given the organization's need to demonstrate added value."[43]

In other words, the terrorist warning machine was fragmented before 9/11 and the president's staff was left to decide what to make of the various terrorist threat reports coming its way. It was something akin to having different voices whispering into different ears. It was a recipe for confusion.

As it evolved, the NCTC is "the only place in the IC [intelligence community] where analysts have the authority and access to fuse a variety of sensitive foreign and domestically acquired data about KSTs [known and suspected terrorists] with other datasets, ranging from financial, travel, immigration, identity, event, seized-media, and IC reporting."[44]

John Brennan, a former daily intelligence briefer to President Clinton, was the first TTIC director. He was the CIA station chief in Saudi Arabia at the time of the 1996 Khobar Tower bombings that killed nineteen Air Force personnel and injured four hundred others. Brennan would later go on to be CIA director under Barack Obama. In 2003, Brennan was asked by then CIA director George Tenet to establish TTIC. In his memoir, Brennan discusses "the challenge of a government 'start-up' " and highlights the task of overcoming the different-voices problem. Among his earliest decisions was the choice to rent a seven-story building near Tysons Corner, Virginia. His new team needed a place to work, and he didn't want them

---

[43] NCTC, *Inside NCTC* (Washington, DC: NCTC, 2021), https://www.dni.gov/files/NCTC/documents/features_documents/InsideNCTC-2021.pdf.

[44] NCTC, *Inside NCTC*.

spread across the greater Washington, DC, area in their home institutions.

Brennan then had to corral his new staff at their new headquarters. He recalls "a rather loud and animated telephone conversation with FBI Director Bob Mueller, in which I pushed back forcefully on his contention that TTIC was a CIA entity and that I was carrying out clandestine CIA organizational objectives. 'You are badly mistaken, Bob,' I practically shouted over the phone. 'The CIA is on the warpath against me for the very same reason your organization is now rebelling against the TTIC model—you guys don't want to share your data. Well, the president says you must, and I am going to continue this fight until TTIC gets access to FBI networks, CIA networks, and any other network critical to our counterterrorism mission.' "[45]

Even in a time of war, changing the government is not easy.

Perhaps the most controversial retooling of the warning machine was the program established to monitor the cell phone and email traffic of terrorist suspects, including persons within the United States. This program was later known as the Terrorist Surveillance Program.

George Tenet recalls, "Prior to 9/11 there was precious little domestic data gathered. We had no systematic capability in place to collect, aggregate, and analyze domestic data in any meaningful way. Domestically, there were few if any analysts. There was no common communication architecture that allowed the effective synthesis of terrorist-related data in the homeland, much less the seamless flow of information to state

---

[45] John Brennan, *Undaunted: My Fight Against America's Enemies, at Home and Abroad* (New York: Celadon Books, 2020).

and local officials in the United States. At the beginning of the twenty-first century, U.S. intelligence officers in Islamabad could not talk to FBI agents in Phoenix."[46]

General Mike Hayden was the National Security Agency director at the time. He would later go on to be CIA director. He explains the various programs in detail in his memoir.[47] For example, he notes, "In covering foreign intelligence targets, it is not uncommon to pick up communications to, from, or about an American. When that happens, NSA [National Security Agency] is allowed to continue to collect and indeed to report the information, but the US identity—unless it is critical to understanding the significance of the intelligence—is obscured, or what we call 'minimized.' The name of the individual, for example, becomes 'US person number one.' "[48]

Hayden and the NSA team designed a program that focused on creating "metadata" that revealed patterns and networks among communications as opposed to individual links. He describes the new program, which was known as "Stellarwind," as "aggregating domestic metadata (the fact of calls to, from, and within the United States) and another that effectively allowed us to quickly intercept the content of international calls, one end of which might be in the United States, if we had reason to believe the call was related to al-Qaeda."[49]

Hayden goes on to say, "When we were fully set up, just because of the way the telecommunications network functions,

---

[46] George Tenet, *At the Center of the Storm* (New York: HarperCollins, 2007), 502.

[47] Michael V. Hayden, *Playing to the Edge* (New York: Penguin Press, 2016).

[48] Hayden, *Playing to the Edge*, 65.

[49] Hayden, *Playing to the Edge*, 67.

we had the theoretical ability to access a significant percentage of the calls entering or leaving the United States."[50]

Stellarwind began as a secret program that ultimately became public—and when it did, many expressed outrage. The legal basis for the program was hotly debated within the Bush administration and by members of Congress. On several occasions, the Bush administration attempted to dissuade the *New York Times* from publishing details of the program. When the story finally made print, it ran under the headline BUSH LETS US SPY ON CALLERS WITHOUT COURTS.[51] Public outrage over the "warrantless wiretapping" did lead to new legal structures governing how the government could gather such information.

Beyond the intelligence community, there was a desire to build new tools for the action machine, to add arrows to the quiver beyond the military. The Treasury Department snapped into action looking for ways to disrupt terrorist financing. Paul O'Neill, Treasury secretary, established a team to deny terrorist organizations the finances they needed to operate. O'Neill gave new life to an old organization with a rather innocuous name—Office of Foreign Assets Control, or OFAC—but it proved to be a powerhouse in finding and disrupting the flow of funding to terrorist organizations. Targeting enemy finances was hardly a new tool, however. It has been an element of statecraft since the dawn of time and was an important tool employed against Japan in the run-up to World War II.[52]

---

[50] Hayden, *Playing to the Edge*, 74.

[51] "Bush Lets US Spy on Callers Without Courts," *New York Times*, December 16, 2005.

[52] Edward S. Miller, *Bankrupting the Enemy: The U.S. Financial Siege of Japan Before Pearl Harbor* (Annapolis, MD: Naval Institute Press, 2007).

Juan Zarate, who served under O'Neill, notes, "After September 11, 2001, we unleashed a counter-terrorist financing campaign that reshaped the very nature of financial warfare. The Treasury Department waged an all-out offensive, using every tool in the toolbox to disrupt, dismantle, and deter the flows of illicit financing around the world."[53]

What followed was the creation of both an intelligence and an action arm at Treasury—intelligence on terrorist financing and relationships, and action to disrupt and degrade the networks. The new tools and authorities for using them were codified in the USA PATRIOT Act, signed by Bush in October 2001. Other partners, including Russia and China, were brought into the fold. The tools themselves had to be tailored to the task. As Zarate recounts, not everything merits a "nuclear option." Treasury analysts were dispatched to the military commands in Europe and the Pacific, as well as the Middle East command that was located in Florida.[54] In time, Treasury created the Office of Intelligence and Analysis, "the first finance ministry in the world to have an arm with an active intelligence function."[55]

As with most powerful governmental tools, it occasionally came in conflict with other terrorist-targeting operations. Zarate recalls a 2002 visit from Cofer Black of the CIA's Counterterrorism Center, who came to alert his Treasury colleagues that naming certain targets could disrupt other potential operations that were being planned. Black cautioned that "outing known terrorist supporters and networks...needed to be

---

[53] Jaun Zarate, *Treasury's Wars* (New York: PublicAffairs, 2013), 7.

[54] Zarate, *Treasury's Wars*, 195.

[55] Zarate, *Treasury's Wars*, 206.

coordinated with the clandestine and covert operations underway around the world."[56]

Part of the new operation was gaining access to SWIFT financial transaction data. SWIFT stands for Society for Worldwide Interbank Financial Telecommunication. It is a financial messaging service that records all member transactions, or as Zarate describes, it is "the switchboard of the international financial system."[57] Prior to 9/11, the US Treasury did not have access to SWIFT data. But access was negotiated in October 2001, which led to the creation of Treasury's Terrorist Financing Tracking Program, which, Zarate recalls, was affectionately known within Treasury circles as Turtle, the opposite of SWIFT.[58]

Treasury's work, in time, pointed to some uncomfortable suspects, including the government of Saudi Arabia and its support for Islamic charities, which included the schools or madrassas that were the feeding grounds for radical Islam. When confronted with the evidence by O'Neill, the Saudis proved willing to cooperate.

The retooling that took place at Treasury in the immediate aftermath of 9/11 would have much wider applicability in the coming years. A more surgical sanctions machine proved invaluable when the United States confronted North Korea, Syria, and Iran over their nuclear weapons programs. It helped disrupt the AQ Khan network in Pakistan that was marketing nuclear technology. It was used to isolate and punish Iraq prior to the Iraq War and helped prosecute the war on drugs. The new weapons were wielded in ways that would deter other

---

[56] Zarate, *Treasury's Wars*, 40–41.

[57] Zarate, *Treasury's Wars*, 50.

[58] Zarate, *Treasury's Wars*, 54.

terrorists, or as Zarate memorably states, it was like "killing the chicken to scare the monkeys."[59]

It was no accident that President Biden called on the Treasury team when he decided to impose sanctions on Russia after its invasion of Ukraine. The mechanisms built over the last two decades enabled the Biden team to swing into action so quickly. The tools were built to constrain not just the Russian economy but also Vladimir Putin and the Russian oligarchs themselves.

Other tools that had been built up quietly over the years by the action machine paid off in significant ways, and in public. The heroics of SEAL Team 6 in the raid against Osama bin Laden have become the stuff of legend. The United States had been hunting bin Laden since well before the 9/11 attacks. He was implicated in a growing list of attacks against the United States since the mid-1990s. It fell to SEAL Team 6 to fly by helicopter into Pakistani airspace and conduct the nighttime raid that led to bin Laden's death on May 2, 2011. This operation was the culmination of nearly a decade-long manhunt that began in the immediate aftermath of the 9/11 attacks. It was during this time that the Defense Department under Donald Rumsfeld began a major expansion and reorientation of the US Special Operations Forces, or SOF.[60] Within two years of 9/11, SOF funding had doubled. By 2008, SOF funding had nearly quadrupled.[61] SOF funding has continued an upward climb to

---

[59] Zarate, *Treasury's Wars*, 220–237.

[60] The National Security Agency was also under the authority of the secretary of defense, though Hayden makes clear that instructions on Stellarwind were coming from the president and were being coordinated closely with the Department of Justice.

[61] FY22 Green Book, https://comptroller.defense.gov/Portals/45/Documents/defbudget/FY2022/FY22_Green_Book.pdf.

this day and represents but one aspect of the investment made to fight terrorist organizations globally.

Soon after the 9/11 attacks, Rumsfeld took additional steps to clarify SOF responsibilities. Rumsfeld saw the terrorist threat as being global, but the US military is organized by region. Missions are assigned to regional commanders, and no single commander is responsible for planning across regions, or theaters, as they are known.[62] The terrorist threat was an amalgam, a vast global network, with connections across regions. Rumsfeld wanted to be sure the people responsible for disrupting and destroying this network had a similar perspective, that they, too, assumed they were operating across a vast global network. He wanted to ensure that, if he directed action in one area, there would be someone responsible watching, warning, and acting in another. He wanted the Special Operations commander to play this role.

The Unified Command Plan, or UCP, is the tool the secretary of defense uses to assign regions and responsibilities to the combatant commands—the part of the military responsible for conducting military operations inside and outside the United States. Geography had been the dominant factor in assigning responsibilities, with two exceptions: US Strategic Command, which is responsible for planning the use of nuclear weapons, is not limited by geography, though it is only responsible for planning against hostile nuclear weapons states; and US Transportation Command is similarly responsible for moving troops and materiel around the globe. Soon after 9/11, Rumsfeld directed

---

[62] The role of the military services is to organize, train, and equip forces according to their assigned roles. For operations, forces are assigned to theater commands, such as US Central Command, or CentCom, which has responsibility for operations across the greater Middle East.

that Special Operations Command be given global planning responsibilities. He was always conscious of the "seams," or boundaries, that divided the commands, and in the pursuit of global terrorist networks, he wanted to ensure there were no seams.[63] He wanted to give his Special Operations commander the authority to pursue terrorist networks where it led them.

Beyond the retooling, the activity of the Defense Department itself told the story of zoom for the two decades that followed 9/11. Within the military, it is common to hear supported and supporting relationships discussed. These terms refer to how forces from one military service or region support others. For much of the Cold War, the US European Command was the supported command, meaning forces arrayed across other regions were directed to support the European Command should war break out in Europe. Similarly, in the event of hostilities, should the focus be on a particular ground battle, then air and naval forces in the vicinity would be supporting the ground commander. Should the focus be on an air attack, then ground and naval forces would be supporting the air commander. Of course, these supported and supporting roles are intended to be fluid over time.

In the immediate aftermath of 9/11, US Central Command, which is responsible for the greater Middle East area and which was established just after the 1979 revolution in Iran, became the supported command. Central Command was first among equals when it came to requests for troops and overall defense resources. Indeed, in a reversal from the Cold War years, forces stationed in Europe and Korea were routinely deployed to

[63] A. R. Hoehn was on Rumsfeld's staff and was responsible for the Unified Command Plan during this time.

Central Command to help with the ongoing fighting in Iraq and Afghanistan. Similarly, naval forces, especially aircraft carriers, were dispatched to Central Command often at the cost of supporting other priorities in Europe and especially the Pacific.

Beyond fighting forces, Central Command had priority for intelligence, surveillance, and reconnaissance forces, too. New capabilities like Global Hawk, a high-altitude surveillance drone, and Predator, a medium-altitude surveillance drone with attack capabilities, were assigned primarily to missions in the greater Middle East area. Use of these assets was necessary in targeting terrorist and insurgent activity in some of the most prominent missions, such as the bin Laden raid. But it also meant they were not available for missions elsewhere.

In the end, it would be an exaggeration to say all eyes and hands were supporting a single region because the warning and action machines operate on a global scale. But it is safe to say that in the aftermath of 9/11 and for nearly two decades after, most of the eyes and most of the hands were devoted to supporting a single regional command.

But as the military got bogged down in Afghanistan and Iraq, it had to do a different kind of retooling. Just as the Army and Marines had to create counterinsurgency doctrine in Vietnam to fight off insurgent forces and pacify the local population, the two services found themselves again writing doctrine for wars that were going poorly. Army Lieutenant General David Petraeus led the effort, with assistance from Marine Lieutenant General James Mattis. They applied the new approach along with a substantial surge in forces that brought just enough stability to key areas to allow for a handoff to the new Iraqi Army.

The military services themselves had to be adapted to this new kind of fighting, especially the Army. The Army had traditionally been organized around large fighting forces known as brigades, divisions, and corps. Several brigades comprised a division, which had its own support forces. Similarly, several divisions comprised a corps, along with sizable support forces like artillery, transportation, and communication. Prior to the wars in Afghanistan and Iraq, the Army organized in corps and divisions. The problem was that the number of corps and divisions was relatively small—four corps and ten divisions in the active or full-time Army. To sustain the demanding schedule of rotating forces in and out of both Afghanistan and Iraq, the Army changed its basic organizational structure to brigades. There were more than thirty brigades in the full-time Army, and this provided a better base of organizational units to rotate in and out of both areas. Even with this change, the Army felt itself coming and going, which led some to declare the Army was stretched thin.[64] The Marines had to make their own adaptations to support continuous deployments.

All this coming and going went on for nearly twenty years, though the most demanding periods were during the surge in Iraq in 2006–2007 and the surge that followed in Afghanistan in 2011.

What this also meant for the Army and the Marines, in particular, though it could be said for the other military services as well, is that they were doing little else. The demands of Afghanistan and Iraq were a full-time commitment not only

---

[64] Lynn E. Davis, J. Michael Polich, William M. Hix, Michael D. Greenberg, Stephen D. Brady, and Ronald E. Sortor, *Stretched Thin: Army Forces for Sustained Operations* (Santa Monica, CA: RAND Corporation, 2005), https://www.rand.org/pubs/monographs/MG362.html.

for the regional command but also for the military services that had to organize, train, and equip forces to support them.

In a very specific illustration of the retooling—and of bureaucracy doing its thing—Secretary of Defense Robert Gates had to sidestep the bureaucracy to purchase a specialized armored vehicle, which would become known as the MRAP (mine-resistant, ambush-protected), that could withstand roadside attacks. Too many American military personnel were being maimed and killed by roadside bombs, first in Iraq, then in Afghanistan. Deployed forces were developing workarounds, like welding armor plates to their vehicles or putting sandbags on floors, but the improvised bombs continued to injure more than the workarounds were protecting. Gates learned the Marines were experimenting with a new vehicle and wanted to learn more. The more he learned, the more he realized he needed to get the vehicle to Iraq, and in significant numbers. Gates could not count on the bureaucracy to get the job done, so he elected to do it on his own—an incredible indictment of the machine. Instead, he directed a team to work with industry to purchase a new vehicle that could withstand the attacks. Many in the Army didn't like it, but once commanders in the field knew they were available, they requested MRAPs by the thousands. There is no question the new vehicle saved American lives. This small but important illustration raises the question of why it took the secretary of defense to intervene to save the lives of American military personnel fighting in a far-off war.[65]

The action machine was adapting to fight the latest wars, but it was not necessarily focused on the right problems.

---

[65] Robert M. Gates, *Duty: Memoirs of a Secretary at War* (New York: Alfred A. Knopf, 2014), 119–126.

In the end, Rumsfeld had this to say about his efforts to retool the action machine to place it on zoom: "I made it a priority of increasing the size, capabilities, equipment, and authorities of the special operations forces. By 2006, we had boosted their funding over 107 percent, doubled the number of recruits, and improved their equipment substantially. I authorized the Special Operations Command (SOCOM) as the lead command for war on terror planning and missions." He noted several other changes to give special forces greater reach and agility. But then he could not resist noting that not all his changes were welcome. "Even though these were historic changes for the armed forces, they were resented by those wedded to the conventional, traditional Army."[66]

What is striking when looking back at the recollections of Bush, Cheney, Rumsfeld, Tenet, and many, many others is not just changes they put in place but also the intense focus, indeed mindset, the 9/11 attacks brought to each of them in their roles.

As George Bush recalls:

> As I listened to my last CIA briefing the morning before President Obama's Inauguration, I reflected on all that happened since 9/11: the red alerts and false alarms, the botulinum toxin we thought would kill us, and the plots we had disrupted. Years had passed, but the threat had not. The terrorists had struck Bali, Jakarta, Riyadh, Istanbul, Madrid, London, Amman, and Mumbai. My

[66] Donald H. Rumsfeld, *Known and Unknown* (New York: Sentinel, 2011), 654.

> morning intelligence reports made clear that they were determined to attack America again.
>
> After the shock of 9/11, there was no legal, military, or political blueprint for confronting a new enemy that rejected all the traditional rules of war. By the time I left office, we had put in place a system of effective counterterrorism programs based on a solid legal and legislative footing.[67]

In other words, Bush had retooled the machine and placed it on zoom, though his claim of a solid legal footing remains controversial.

Dick Cheney concludes his memoir by noting "nothing was more important than having kept the nation safe after the devastating attacks of 9/11." Cheney goes on at great lengths to defend the Bush administration's record, but he makes no apology for the decisions he thought were essential to avoid a repeat of the 9/11 horrors, including the use of harsh interrogation as an intelligence collecting technique. In reflecting about 9/11, Cheney recalls:

> I thought about the fact that the city of Washington had come under attack in 1814 at the hands of the British. Now, 187 years later, al Qaeda had demonstrated they could deliver a devastating blow to the heart of American economic and military power. On this day, all our assumptions about our own security had changed. It was a fundamental shift.... We were in a new era and needed an entirely new strategy to keep America secure. The first

[67] George W. Bush, *Decision Points* (New York: Crown, 2010), 179.

> war of the twenty-first century wouldn't simply be a conflict of nation against nation, army against army. It would be first and foremost a war against terrorists who operated in the shadows, feared no deterrent, and would use any weapon they could get their hands on to destroy us.[68]

George Tenet, Bush's CIA director, expresses a nearly identical view: "Terrorism is the stuff of everyday nightmares. But the added specter of a nuclear-capable terrorist group is something that, more than anything else, causes me sleepless nights. Marry the right few individuals with the necessary materiel, and you could have a single attack that could kill more people than all the previous terrorist attacks in history. Intelligence has established beyond any reasonable doubt the intent of al Qaeda is to do precisely this."[69]

With damning clarity, not only did the zoom-like focus on terrorism allow other threats to proliferate and rise without proper attention, but also values sacred to this nation were set aside in the rush to find, fix, seize, or kill terrorists and disrupt their plots. "Enhanced interrogation" became a euphemism for what most reasonable people would call torture. Black sites operated without typical oversight, as the name makes clear. The abuse of Iraqi prisoners by their American military jailers at Abu Ghraib stained this nation's reputation and became a recruitment poster for extremism.

Still, 9/11 was the defining moment of the Bush presidency. Plans for the attack were well underway before Bush was elected. When the attack arrived, it was evident the nation

---

[68] Dick Cheney, *In My Time* (New York: Threshold, 2011), 10.

[69] Tenet, *At the Center of the Storm*, 503.

was not prepared for what would follow. If nothing else was to be achieved during the Bush presidency, another terrorist attack on the scale of 9/11, or worse, had to be prevented or disrupted. If it required placing the gargantuan warning and action machines on zoom, the administration was determined to do so. The Bush team was focused on bringing Congress along each step of the way, but they would not await congressional action to make the decisions they thought necessary to protect the nation.

Looking back over the past twenty years, it is difficult to express the intensity of the moment. Bush and his advisers were dealing with an attack to the heart of America's political, economic, and military power. What they saw in the carnage was the reality of the moment—nearly three thousand Americans killed in a single attack—but also the sense that far, far worse might yet come. The thought of a biological or nuclear attack that might kill many thousands or even millions of Americans forced them into action and had a powerful effect that would guide their decisions over the remaining years of the Bush presidency. It would continue through the Obama years even though Bush's successor was committed to returning the nation to greater normalcy. The idea that a group operating in the shadows could bring devastating harm, as Cheney expressed, had a way of pushing other important needs into the future. Once the warning and action machines were put on zoom, it proved difficult to change the focus, even for those, like Obama, who campaigned on the promise to do so.

As Facebook and other social media became so prominent in the mid-2000s, it took on more and more importance for the intelligence community. Open-source information has always been a target for intelligence collection. Old photos of Soviet analysts reading *Pravda* come to mind. For decades the CIA funded the Foreign Broadcast Information Service, known more commonly by its initials FBIS and pronounced "fibis," for its translations of foreign newspapers and television broadcasts. These were in high demand among not only government analysts but also think tank and university researchers.

But social media presented something new. It offered a way to tap into what real people were thinking and doing in real time. It offered ways to gauge not just elite opinion but also popular sentiment. Andy Roberts was put in charge of the effort at the Defense Intelligence Agency. He described social media as a new form of open-source intelligence, or OSINT. He explained to us, "A lot of younger officers were starting to realize the power behind it, especially on the social media side. And within I'd say about a year, in talking to the [combatant] commands, talking to the services, you know, folks within the intelligence community writ large, [we] realize[d] that we really had to establish a foundation."

But realizing a need and making it real are two different things. Roberts described the tug and pull for resources within his own organization. "That, no kidding, was my biggest problem until probably the very end, that the prevailing wisdom really was this is open-source. It doesn't cost money. You know, I can go home and Google this. I really did have a lot of seniors in the beginning tell me that, and it wasn't until we did an active campaign of trying to sensitize folks that this isn't the old OSINT."

He went on to describe the talent, the tools, and the creation of a new intelligence discipline. He described recruiting a new workforce that was not interested in wearing ties to work. He talked about the need to use machine learning and artificial intelligence to sort the wheat from the chaff.

In helping create a new intelligence discipline in a community where there are rivalries and naysayers, he described his challenge like this: "It's literally like looking through multiple haystacks to find what you want to look for."

Everybody is in search of the elusive needle.

Yet, another innovation of the 9/11 era was use of the internet to tap into the wisdom of crowds. Just as open-source information proved to be an invaluable addition to the intelligence treasure trove, others in the intelligence community were experimenting with the idea that informed observers might be just as good at peering into the future as experts, or perhaps even better.[70] The Intelligence Advanced Research Projects Activity, better known by its initials IARPA, is one place where these experiments were being run.

IARPA came about as part of the post-9/11 reforms. The idea was that in the information age, the intelligence community needed the same technological prowess that the Defense Department had so long prided itself on. The Defense Department had

[70] An interesting window on expert advice is found in Dick Cheney's memoir *In My Time*. Cheney describes meeting with US ambassador Charles Freeman just before conferring with Saudi King Fahd prior to the First Gulf War. Cheney recalls Freeman telling him, "You have to be cautious. . . . If you are too aggressive or talk about too large a force, you will scare the Saudis and they won't commit." Just before meeting with Fahd, Prince Bandar, Saudi ambassador to the United States, told him, "It's very important you demonstrate to the king that you are serious." Cheney goes on to say, "He wanted me to make sure the king knew we would commit a large force and do it fast" (Cheney, *In My Time*, 189–190).

the Defense Advanced Research Projects Agency (DARPA) that was famous for taking risks and producing breakthrough technologies, including the ARPAnet, the predecessor to the modern internet. The hope was that IARPA could do similar things for the intelligence community.

IARPA's attempts at improving prediction methods and applying big-data analytics to other problems have produced fascinating successes—including cases where mathematicians and data engineers proved better at prediction than intelligence experts.

To keep analysts primed and nourished, IARPA's leadership would regularly sponsor "prediction competitions"—offering free pizza as the prize.

IARPA also sponsored a massive, multiyear competition aimed at testing new kinds of prediction techniques. The details are nicely recounted in Philip Tetlock's book *Superforecasting: The Art and Science of Prediction*. It was Tetlock who expressed the now famous line about prediction: "The average expert was roughly as accurate as a dart-throwing chimpanzee." There is a much longer story behind Tetlock's chimpanzee line, though it is safe to summarize that many in the prediction business do little better than average. It is a similar logic that leads investors to funds that represent large segments of the market rather than individual stocks.[71]

Even as IARPA was examining other threats and the United States marked twenty years since the attacks of 9/11, lingering military commitments to the wars in Afghanistan and Iraq are perfect case studies for failures of the warning–action relationship.

---

[71] Philip Tetlock and Dan Gardner, *Superforecasting: The Art and Science of Prediction* (New York: Crown, 2015).

The rise of ISIS, we now understand clearly, if belatedly, was a threat of monstrous lethality. Initially dismissed by President Obama as the "junior varsity" team of terrorism, ISIS rose in 2014 to seize and rule a swath of Iraq and Syria the size of the United Kingdom and inspired deadly attacks around the world—including lethal shootings by ISIS adherents inside the United States.[72] It remains a menace to this day.

How did we miss this, and not act? Was it a problem of warning or of action?

In this case, it was both, with responsibility to be shared by President Obama and his senior national security team, military commanders at the Pentagon and in the Middle East, and officers on the ground in Iraq.

Several senior government and military officials, now retired, told us of intelligence community assessments that accurately anticipated the rise of a potent terrorist organization as successor to al-Qaeda in Iraq.[73] However, these officials acknowledged that the assessments—as is often the case—were written in careful, measured, on the one hand/on the other language that the impact and "end state" were open to interpretation.

Obama's senior advisers knew the president wanted out of Iraq, and badly. So, they took these classified assessments, emphasized the uncertainty of the predictions, and assured the

---

[72] David Remnick, "Going the Distance," *New Yorker*, January 19, 2014, https://www.newyorker.com/magazine/2014/01/27/going-the-distance-david-remnick.

[73] Interviews with retired senior government officials over the summer of 2019. They spoke on condition of anonymity to describe assessments they had read and briefings they received concerning intelligence on ISIS.

president "It won't happen," according to one retired official who sat in on the discussions.

Contrast that with a senior general, now retired, who also sat in on top-level planning sessions for the military mission in Iraq and received the same intelligence briefings. Having built a successful career betting on worst-case scenarios, this general told us, he and several others deeply read and understood warnings about the rise of a shadowy new adversary in Iraq as a successor to al-Qaeda.[74]

But this senior official said he also believed that the billions of dollars spent to train and equip Iraqi Security Forces, under the guidance of their American military advisers on the ground, would hold back a flood of new bloodletting.

And on the ground in Iraq? We do not know of anyone who served in Iraq who had confidence in the Iraqi army. No midlevel military officer wants to tell the boss things are going badly. So, whether willfully or as acts of aspirational good faith, the American military trainers seemed to have never shared the depth of their concerns about the fragility of the Iraqi Security Forces. On the ground in Iraq, the trainers saw the lack of morale, the widespread corruption and cronyism, and hardening factionalism between various tribes and religious groups that made up the Iraqi Security Forces. Yes, these officers filed regular Situation Reports, but nobody raised a red flag up the chain of command to say clearly that these Iraqi forces will collapse under attack from an organized anti-government militant force.

Which is exactly what happened. Adding to the error in failing to assess the rise of ISIS was one of personnel. As the

---

[74] Interview with retired senior general on June 18, 2019. The general insisted on anonymity to discuss intelligence on ISIS.

American force levels continued to dwindle, there were fewer and fewer soldiers to serve as "sensors" on the ground in Iraq. And the vaunted surveillance drones could not make up for it.

Just one Predator surveillance flight a week was available to watch those contested areas of northern and western Iraq, according to officials involved in the mission, down from dozens during the height of the war.[75] American intelligence was unable to see the carefully calculated efforts to undermine civil authority carried out by a nascent ISIS in a four-year, ruthless campaign that murdered thousands of local Iraqi law enforcement officers and military officials, as well as assassinated pro-Baghdad tribal and civic leaders. Important municipal services and infrastructure were destroyed by car bombings and suicide attacks until the population was cowed into submission.

Yes, the intelligence community warned of the rise of ISIS. But what senior leaders heard was an uncertain trumpet...and death and disaster followed.

And the inability to accept accurate descriptions of corruption and failed efforts in Afghanistan were failures just as significant.

The horrific collapse of American policy in Afghanistan in the summer of 2021—and the collapse of the Afghan security forces and government sponsored by the United States with trillions of dollars over twenty years—followed a similar path, but with one important difference. The only people who were unaware of the fragility of Afghan national security and defense forces were those who were not paying attention.

The warning community is well aware of the myth of Cassandra, the Trojan priestess blessed by the gods with the power

---

[75] Interviews with intelligence and military officials over the summer of 2019.

to clearly see the future—but cursed by the gods to never be believed.

Cassandra, meet John Sopko, the special inspector general for Afghan reconstruction (SIGAR), who is the mythic priestess's real-life Washington equivalent. His mission was to serve as watchdog over the hundreds of billions of dollars spent in Afghanistan reconstruction, in particular on security and defense forces.

Sopko's quarterly reports to Congress were aggressive, unwavering, clear-eyed. Anyone who paid even the slightest attention to the SIGAR assessments could not have been surprised that the Afghan national security and defense forces melted quickly after President Joe Biden's announcement of a full US withdrawal from the country in the summer of 2021, a collapse of government order that cleared the way for the Taliban to sweep to power and rout the dreams of millions of ordinary Afghans.

"After all the money, $86 billion and 20 years, why did we see such poor results?" Sopko asked. "You really shouldn't be surprised, if you've been reading our reports. For at least over the nine years that I've been there, we've been highlighting problems with our train, advise and assist mission with the Afghan military."[76]

The military has a concept called "Commander's Intent." Much like the country music axiom, "If you don't know where you're going, any road will do," Commander's Intent should

---

[76] John Sopko, "John F. Sopko, Special Inspector General for Afghanistan Reconstruction" (transcript), Defense Writers Group, Project for Media & National Security, George Washington School of Media and Public Affairs, July 29, 2021, https://cpb-us-e1.wpmucdn.com/blogs.gwu.edu/dist/2/672/files/2018/02/DWG-Sopko-210729.pdf.

set clearly defined milestones to reach a desired goal. Yet Sopko's reports criticized the American military for "moving the goal posts every time we took a look at the assessment tools. Our U.S. military would change the goal posts and say, 'Oh no, no, that's not the test you want to do.'" His reports raised significant questions about the sustainability of the high-tech hardware the United States gave to Afghan troops, many of whom were barely literate. And his reports warned of the lack of sustainable logistics to support Afghan forces, in particular the air force. And, of course, they identified the corruption that robbed soldiers on the ground of fuel, food, bullets—and the will to fight the Taliban.

"Napoleon said during the 1800s that an army moves on its stomach," Sopko told us. "And that is so true. And if you expect the Afghan military to win the hearts and the minds of the Afghan people, you have to win the hearts and minds of the Afghan military. If you don't pay them, you don't feed them, you don't support them, you don't give benefits to widows and orphans on a regular basis, you don't have Medevac capabilities, then the average Afghan soldier is saying, 'What the heck am I dying for?'"

Sopko chose two words to explain why policymakers in Washington did not heed warnings about the hollow Afghan security forces: hubris and mendacity.

> One is hubris, that we can somehow take a country that was desolate in 2001 and turn it into little Norway in that timeframe. The other thing is mendacity. We overexaggerated—our generals did, our ambassadors did, all of our officials did—to Congress and the

> American people, about we're just turning the corner. We're about ready to turn the corner. We can give you chapter and verse about how many of our generals talked about being just about ready to win. Well, we turned the corner so much, we did 360 degrees. We were like a top.
>
> We have to be honest with ourselves and we have to be honest with the American people who pay for this. Not only in money but also in blood and treasure. It's a complicated issue, but put all of them together and this is why we got what we got.

The ongoing pandemic, the threat of never-ending fires in the American West, Russia's war with Ukraine, and China's increasingly aggressive moves toward Taiwan have awakened America from its twenty-plus-year focus on terrorism and the Middle East. The threat posed by the 9/11 attacks was consequential but never existential. And Osama bin Laden, the terror mastermind behind those attacks, was killed in May 2011.

That was more than a dozen years ago. In the meantime, a new age of danger has emerged. The threats include new weapons and new threats, to be sure, just as they include traditional Great Power rivalries and new superpowers. The high-end risk was defined with frightening clarity in late 2022 by the four-star admiral who commands America's nuclear arsenal.

"This Ukraine crisis that we're in right now, this is just the warm-up," said Admiral Charles Richard, commander of

US Strategic Command.[77] "The big one is coming. And it isn't going to be very long before we're going to get tested in ways that we haven't been tested in a long time. We have to do some rapid, fundamental change in the way we approach the defense of this nation." Assessing the threat from Russia, he added, "The current situation is vividly illuminating what nuclear coercion looks like and how you or how you don't stand up to that."

He then refocused from Moscow's aggression to the rising power in Beijing. "As I assess our level of deterrence against China, the ship is slowly sinking. It is sinking slowly, but it is sinking as, fundamentally, they are putting capability in the field faster than we are," he said. As those curves keep going, Admiral Richard warned, the balance with Beijing no longer is decided by "how good our commanders are, or how good our forces are. We're not going to have enough of them. And that is a very near-term problem."

---

[77] C. Todd Lopez, "Stratcom Commander Says U.S. Should Look to 1950s to Regain Competitive Edge," US Department of Defense News, November 3, 2022, https://www.defense.gov/News/News-Stories/Article/Article/3209416/stratcom-commander-says-us-should-look-to-1950s-to-regain-competitive-edge/.

PART II
GREAT POWER RIVALRIES AND THE NEW SUPERPOWERS

CHAPTER 3

CHINA

Time Is on Our Side (Until It Isn't)

For over three decades, specialists in the United States have debated the meaning and consequences of China's emergence as a leading economic and military power. For the US business community, China's development meant opportunities for business in the form of new markets and the opportunity to produce goods at much lower cost. For consumers, it meant cheaper goods and lots of them, anything from household products at the Dollar Store or Big Lots to big-screen TVs at Best Buy or the local Walmart. For more specialized communities interested in human development, China's emergence meant that more than a billion people would have the chance to emerge from poverty and enter the middle class. But for military specialists, it was a much darker sign. China's emergence potentially meant that a new rival might evolve that could challenge US strategic and commercial interests in the Pacific and threaten US allies.

For much of the 1990s and 2000s, these various viewpoints of China and its development ran roughly in parallel. The predominant business attitude—along with many in policy and governing circles—was that China's economic development would be accompanied by a change in politics. The grip of the

Communist Party would change over time as China became a more "regular" nation, a contributor as well as a benefiter of the larger international system. Sometimes a sentiment was expressed that went something like this: that which can't last won't last, meaning the Chinese Communist Party (CCP) could not maintain control atop a burgeoning economic megalith. This line of thinking led to the development of an attitude that time was on our side because China's development would lead the CCP to value the benefits of the larger international economic system that was bringing wealth and prosperity to China. A growing chorus within American and European circles called for China to become a responsible stakeholder—that is, to be a constructive player within the existing international system.

At the same time, a large number of national security analysts were watching China's military modernization efforts with increasing alarm—if China fielded the full array of military capabilities it was developing, especially its missiles and targeting capabilities, many wondered if the US military and its regional partners would have the requisite capabilities to respond in the event of an attack on US facilities or a US ally. Still, with the wars in Afghanistan and Iraq demanding constant attention, the debate about China's growing military prowess played out in the background largely among specialists and had to compete for the attention of top policy officials.

The parallel viewpoints—China as a place to do business and China as a growing threat—converged soon after Xi Jinping came to power as the leader of the Chinese Communist Party in 2012. It was not long after Xi's rise that he began talking of two competing international systems and not one, of China's growing dominance on the world scene, and how China would flex its military might if need be. The business community became more wary, and the military community began speaking with more of a common voice about the nature of China's military prowess—military might that was more real than imagined.

This tension was well captured in the standing US policy toward Taiwan, which had been in place for decades. Known as "strategic ambiguity," the policy was supportive of Taiwan's growing democratic traditions, but it was cautious about where these might lead if it meant Taiwan would one day declare independence. Since the time of Jimmy Carter, the United States has maintained a "one China" policy—what some would call a one-country, two-system approach. Should Taiwan's path toward democracy also lead it toward independence, the United States maintained the right to be "ambiguous" about whether it would come to Taiwan's defense. Alternatively, should mainland China seek to bring Taiwan to heel through force, the United States retained the option to come to Taiwan's defense. Of course, policymakers understood the complexity and potentially difficult choices that filled the space between a Taiwan declaration of independence and an unprovoked attack from the mainland. This policy space was filled with ambiguity, indeed.[1]

---

[1] Among the complexities that have vexed the policy community are the many variations of Taiwan moving toward independence short of an outright declaration. Examples include establishing separate diplomatic relationships, seeking participation in the United Nations and other international organizations, developing and deepening commercial and cultural ties, and so forth. Similarly, China need not necessarily orchestrate an invasion of Taiwan to prompt a military response. It could interfere with commercial and fishing traffic in the Taiwan Strait, it could hold large-scale military exercises that demonstrate its ability to invade short of an actual invasion, it could declare the water and airspace surrounding Taiwan to be within China's extended territory, and so forth. As it is, China routinely dispatches warships to the areas surrounding Taiwan and flies military aircraft into the airspace surrounding Taiwan.

Against this backdrop, in 2021 General Mark Milley, chairman of the Joint Chiefs of Staff, spoke out after China tested a new missile that could fly at unprecedented speed and reach targets around the globe, potentially with nuclear weapons. The appearance of this weapon came as something of a shock to the military and policy communities. Granted, China had made great strides in its military technological capabilities over the last two decades. But this seemed different. A major step forward, and a worrisome one at that.

"I don't know if it's quite a Sputnik moment, but I think it's very close to that," General Milley told David Rubenstein, the billionaire and philanthropist who conducts an interview show on Bloomberg Television. The tests, Milley said, were a "very significant technological event" that "has all of our attention."[2] To other observers, General Milley's reaction was more of a WTF or, more politely, a what the heck moment. China's interest in hypersonic weapons has been known for years.

Indeed, this was a rather extraordinary statement given that the US military and the intelligence community have been watching China's hypersonic missile programs—missiles that travel at more than five times the speed of sound—for years. Still, it was jaw dropping, not only to see the test, but also to see it apparently succeed.

What is perhaps most surprising about this development is the way China could employ the new weapon should it prove to be viable. A more traditional ballistic missile is fired from

---

[2] As reported in the *New York Times*: David E. Sanger and William J. Broad, "China's Weapon Tests Close to a 'Sputnik Moment,' U.S. General Says," *New York Times*, October 27, 2021, https://www.nytimes.com/2021/10/27/us/politics/china-hypersonic-missile.html.

point A to point B. It flies in a long arc, and its trajectory can be determined almost immediately after launch. The United States has deployed satellites to detect such launches to warn of a pending missile attack. The flight path of a missile from China or Russia typically is over the North Pole, so the array of sensors and limited defenses against a missile attack—put in place initially to protect against a possible North Korean attack—are also positioned to identify and intercept missiles flying over the North Pole. With this new missile, China could decide to fly it over the South Pole where there are few sensors and no effective defenses. This would demonstrate the CCP's ability to reach distant targets without flying through the teeth of the American missile defense network. In plainspeak: if China wanted to attack, now it could possibly hit America with missiles that might go relatively undetected.

General John Hyten was vice chairman of the Joint Chiefs of Staff at the time. He tried to put China's hypersonic missile test in context. Hyten was finishing his time at the Pentagon and was frustrated by the slowness of Pentagon processes. To Hyten, it was not just what China had achieved but also the commitment China made to the larger effort. "I think in the last five years, maybe longer, the United States has done nine hypersonic tests. In the same time, I can't give you the exact number because that would be classified...the Chinese have done hundreds. Single digits versus hundreds is not a good place."[3]

---

[3] John E. Hyten, "General John E. Hyten, Vice Chairman of the Joint Chiefs of Staff" (transcript), Defense Writers Group, Project for Media & National Security, George Washington School of Media and Public Affairs, October 28, 2021, https://nationalsecuritymedia.gwu.edu/project/general-john-e-hyten-vice-chairman-of-the-joint-chiefs-of-staff/.

Surprised and slow. As General Hyten noted, not a good place.

Hyten was worried about China's missile programs and its overall military modernization effort. "So you have to worry about Russia in the near term, but calling China a pacing threat is a useful term because the pace at which China is moving is stunning. The pace they're moving and the trajectory that they're on will surpass Russia and the United States if we don't do something to change it. It will happen."[4]

But why did this come as a surprise? Had the United States missed an important warning on China and failed to take the required action?

Andrew Marshall was a product of the first Sputnik moment. He was trained as an economist at the University of Chicago and worked at RAND, a Santa Monica, California, research organization, in the early days of the Cold War. Marshall, part of the cohort that author Fred Kaplan called "the Wizards of Armageddon," was among a very select group of people who helped shape early Cold War policy. Their goal was to avoid nuclear war, and they had to create an entirely new discipline to do so. It meant developing new ideas and applying techniques to the most urgent national security threat of the time—nuclear war. Never before in history did two countries have the ability to destroy each other in a brief exchange of weapons. It was an entirely new problem that required an entirely new kind of thinking. Among the ideas they developed was the application of game theory to national security questions. Game theory had

[4] Hyten, "General John E. Hyten" (transcript).

been around for decades in the field of economics, but it had not been applied to national security problems. This cohort produced new games of logic like the "prisoner's dilemma," which is taught in graduate programs across the country to this day.[5] Ideas like the prisoner's dilemma were the product of bringing top minds to some of the world's hardest problems.

Marshall and his colleagues gathered at a unique time in US history. The major universities all had research departments, but few, if any, institutions gathered top talent and worked across disciplines. At RAND, economists worked with engineers. Statisticians worked with logisticians. Historians worked with biologists and anthropologists. When university professors were not teaching in the summer, they would gather with their friends and colleagues in Southern California just a block from the famous Muscle Beach. This was where new ideas came alive—not just ideas about how to avoid nuclear war but powerful ideas that took hold in the social sciences—game theory, systems theory, and a whole lot more.

Many years later, Marshall had become famous as a renowned Pentagon strategist and futurist who directed a little-known but highly respected organization called the Office of Net Assessment.[6] More to the point, he was known as "the Yoda of the Pentagon" among his admirers. True to his

---

[5] The prisoner's dilemma highlights how two rational parties can act in ways that are not in their best interests because they lack important information on how the other party might act—as would happen with two prisoners, both plotting—but separated by bars and walls. This type of theoretical thinking helped guide decision-making during the Cold War.

[6] This section benefits from A. Hoehn's knowledge from working at the Pentagon from 1989 to 2004, interview material with A. Marshall, July 2018, and A. Krepinevich and B. Watts biography of A. Marshall, *The Last Warrior* (New York: Basic Books, 2015).

nickname, Marshall saw what was coming. In the early 1990s, he sponsored a series of wargames on how a modern Chinese military could frustrate US military plans to defend Taiwan in the Pacific. The outcomes were disturbing, even shocking. Time and time again, a modern Chinese military beat the United States at its own game. The strategies the wargamers developed created a whole new language. Expressions like "anti-access" and "area denial" became more and more prominent among defense insiders. Game players quickly recognized that they could adopt military technology—technology that was well within the reach of China's military and industry—that helped the Chinese military make the immediate territory surrounding China something of a "keep away" zone for other military forces. The combination of long-range sensors and highly accurate missiles capable of targeting runways across the region could stop in their tracks US military forces trying to deploy to the region. Without nearby runways or aircraft carriers in range to operate attack aircraft, US military forces would be rendered largely obsolete. That is because the United States depends upon its aircraft—from both the Navy and the Air Force—to defend against attacks on its allies.

In time, others replicated Marshall's results. By the late 1990s, it was generally understood that China had a winning hand to play in the future if it chose to do so. Much of the attention focused on defending against an attack on Taiwan, but the same capabilities that would allow China to invade and secure Taiwan also gave it the ability to exert larger regional dominance. This led Pentagon leaders to call for a transformation of US military forces. Indeed, this was the outside critique of the 1997 National Defense Panel that titled its report "Transforming Defense." Some military commanders tried pushing their own military services into action. For more than two and a half decades, there were statements and pronouncements, working groups on topics like "Air-Sea Battle," and expressions like "pivot," but not a coherent response. With other things going on, particularly the wars in the Middle East, it was often said that time was on our side. Until it wasn't.

China, for its part, felt the sting of embarrassment in the aftermath of the Taiwan Strait Crisis of 1995 and 1996. In response to loose talk of independence among Taiwan's leaders, China initiated a set of missile tests, firing to one side of Taiwan, then the other. William Perry was secretary of defense at the time. He was an old artillery officer, and he knew what it meant to bracket a target. If you could fire to one side and then the other, it also meant you could choose to hit targets in the middle. The missile tests were meant as a warning to Taiwan's leaders—toe the line or risk punishment. In response, the United States dispatched two aircraft carriers to the region. Only one actually arrived, but that was enough to signal that the United States was serious about defending its interests and that China needed to stand down. China was now the one being told to toe the line. China's Communist Party and military were deeply embarrassed by the episode, both in the failure to rein in Taiwan's wayward leadership and in the very visible and provocative US response. In dispatching its aircraft carriers, the United States pressed what CIA China analyst John Culver called the "America wins" button.[7] China's air force was no competition for American airpower at the time. Merely

---

[7] As captured in "China Expert John Culver on Beijing's Military Prowess," *Intelligence Matters* podcast with Michael Morell, June 30, 2021, https://podcasts.apple.com/us/podcast/china-expert-john-culver-on-beijings-military-prowess/id1286906615?i=1000527316077.

showing up meant America wins. China backed off, but its leaders were determined to never again be unprepared should a Taiwan Straits crisis happen again. This resulted in an assertive diplomatic campaign to separate Taiwan from its partners and the start of aggressive military modernization that America's China watchers trace directly to China's embarrassment over Taiwan. That modernization effort continues to this day. From the CCP's vantage point, time was on their side. China was intent upon modernizing its military, and it would take the time needed to get the job done. China also knew that the United States, in the aftermath of 9/11 and amid the wars in the Middle East, was preoccupied with other problems.

For the next twenty years there was a lot of talk about China in America's military circles, but not a lot of action. It wasn't until James Mattis became secretary of defense in 2017 that China's emerging military prowess became the focal point of US defense policy, even though previous Pentagon chiefs had given lip service to a pivot to China. Jim Mattis, a four-star Marine Corps general who'd made his name in the Middle East, realized something needed to be done.

In December 2018, Mattis unveiled a new strategy in a speech at the Reagan Library that grabbed the attention of his listeners. At the time, he was enjoying his position at the top of the military world. After singing the national anthem, he joked in his opening remarks that his lack of musical talents led him to a career as a Marine infantryman. The light tone of his remarks, this from a man known to some as "the warrior monk," gave a sense he was reveling in the moment.

Then Mattis turned serious. He was hired by President Trump to shake things up in the Pentagon, and he was there to do so in ways many might not have expected. His speech was signaling a major change in direction. He and his team at the Pentagon had been working on a new National Defense Strategy for the past year that would set the course for spending over $3.5 trillion over the next five years.

Mattis leaned in; he wanted the audience to understand that the profound lessons of the past were not lost on him. He said forcefully:

> History is unconfused as to what happens when a democracy permits its strength to wane. We see it in our own history. Osan, Korea, 1950, soldiers from Task Force Smith went into battle against enemy tanks carrying obsolete bazookas, incapable of knocking out their targets. We might believe this could never happen in our time. But if the same America that had defeated the Third Reich in World War II could forget in just five years the hard-learned lessons of Anzio, Normandy and the Bulge, so can we in our generation.

What Mattis was announcing was that his country, his department, his military, his Marine Corps were not ready for the threats that were mounting from China and Russia and that it was time to change direction. He was announcing a shift in strategy to focus on threats emanating from what the Pentagon now calls Great Power Competition, or sometimes GPC, for short. He was sending a clear message that it was time for a change in direction. He was challenging not just the military but also the country as a whole—the worlds of industry and finance as well as the military.

Lest anyone doubt Mattis's confidence in his ability to turn the ship of state, he had this to say in conclusion: "We are Americans. We are not spectators in the arc of history. We make history."

Beyond the friendly forum and friendly reception, Mattis was warning the wider national security community that the United States was squandering, indeed perhaps had already squandered, its lead in military technology and was in no shape to compete against other Great Powers. Though others had made similar points, Mattis was the first secretary of defense to say so publicly. He would remain secretary of defense for a little over another year until a dispute with President Trump over Syria policy led to his resignation. Mattis's successor, Mark Esper, would sustain the priorities that Mattis put in motion.

It's not that some hadn't tried earlier. More than fifteen years earlier, Richard Danzig was looking into the future. Danzig was secretary of the Navy in the last years of the Clinton administration. He was undersecretary of the Navy just before that. Danzig is cerebral, very cerebral. The third floor of his Washington, DC, house is a study filled with books. Light streams in from both ends of the room, and he has a wooden office chair where he thinks, converses, and writes. He thinks a lot. He has a Yale law degree and an Oxford PhD. He clerked for Justice Byron White after he graduated from Yale.

Danzig has a sweeping policy mind and a lawyer's bent for details. He has a way of getting to the heart of an issue in seconds, where it might take others minutes or hours. Staying ahead of him, to say nothing of keeping up with him, is no easy task. One of the people who did a good job of this was Bob Work, who was a Marine Corps colonel and Danzig's military assistant. He would later go on to be deputy defense secretary as we will see.

While Danzig was secretary of the Navy, he wrote a paper on future security risks. He called it *The Big Three: Our*

Greatest Security Risks and How to Address Them.[8] It is hard to imagine a military service secretary writing while leading, but it is in the nature of who he is. He has a penchant for looking around corners. In the paper, Danzig reflects on the demands of the time. He notes, "Beyond the present lie other, probably even more important, longer-term issues."[9] China was one of them. Like Andy Marshall, Danzig was calling for a strategy of dissuasion "with the aim of discouraging military competition with us."[10] He was calling attention to the need to focus on the future. "Within the Department of Defense, the present is comfortable and the past is misleading. Nothing seduces like success: why change when your existing method of business has just proved itself a winner in a prolonged contest?"[11] Danzig's caution reads today more like a hedging strategy than a competitive one. He saw upside in integrating Russia and China into the global economy and adopting a dissuasion strategy to help convince potential rivals there was no need to compete. He uses the example of the US Navy protecting oil supplies from the Persian Gulf. To the extent the supply lines are assured, why would another competitor need to build a navy to do the same thing.

Years later in DC, Danzig went on to tell us how he tried to engineer a more dedicated focus on China in the late 1990s.

---

[8] Richard Danzig, *The Big Three: Our Greatest Security Risks and How to Address Them* (Washington, DC: NDU Press, 1999). The other two were "traumatic attacks aimed at sowing anxiety, despair, disruption, and confusion" and "erosion of domestic support" such that the country "may become so indifferent to or, worse, alienated from its military and foreign policy institutions."

[9] Danzig, *Big Three*, 3.

[10] Danzig, *Big Three*, 28.

[11] Danzig, *Big Three*, 31.

He did it in typical Danzig fashion. Find interesting, thoughtful people and get them thinking about his problem.

> I organized behind the scenes in the government a thing called the China group. And we used to meet irregularly on Saturdays, and it became the place to be like every six weeks or something to talk about China. It was really quite wonderful.... It led to a decision that I would go to China in October of 2000. And this was sort of the centerpiece to kind of lay the basis. There had been an embargo on trips to China, people hadn't gone, etc. The Chinese accepted this. I think the Chinese had a theory that [Al] Gore would be elected and that I would be secretary of defense and therefore they needed to [get to know his key advisers], it was a good investment....
>
> I made a heroic effort to get on China time before I left so I could be more capable. I arrived and had dinner with the head of the East fleet, which deals with Taiwan. It was really quite good. I went to sleep, three a.m., and was awakened. The [USS] *Cole* had been bombed. And I decided I had to come back.[12]

An early manifestation of zoom. "So this is an example of how terrorism [overshadowed other threats]. Even as focused as I think I was on my China stuff, I was very aware of some terrorist issues, but then they kind of overtook me. It's a physical, tangible example."[13]

---

[12] Richard Danzig (former secretary of the Navy), interview with authors, March 2022.

[13] Richard Danzig, interview with authors, March 2022.

Soon after, Al Gore lost the election. A new secretary of the Navy would be appointed, the 9/11 attacks would follow, and the United States would spend the next two decades in the Middle East.

Donald Rumsfeld would serve as secretary of defense for the newly elected George W. Bush. Bush was voted into office on a commitment to transforming the military into a high-tech force. He campaigned on the idea of skipping a generation of military technology to ensure the United States was well positioned to maintain its lead. Bush and his team had their eyes on how China was emerging as a major economic and military power, though they would not say it at the time out of concern for offending top leaders of industry who still viewed China as a major market.

But when terrorists attacked the country on 9/11, Bush and his team soon found themselves fighting two unconventional wars in the greater Middle East—one in Afghanistan, viewed as a war of necessity; one in Iraq, viewed as a war of choice. The official view was that there was time to stabilize the Middle East and then focus on China, that a competition in military technology was coming but was not here yet. The nation could finish solving the problems of the day, then turn to the problems of tomorrow. Robert M. Gates, Bush's second secretary of defense and Obama's first, publicly chided those top generals who seemed too preoccupied with future wars, what Gates called future-war-itis, and not sufficiently attentive to the current ones. In fact, some argue he fired the Air Force secretary and chief of staff partly on these grounds, though he dismisses the point in his memoir. But no one, not Bush, Obama, or Trump, imagined that wars that began in Afghanistan in 2001 and Iraq first in 1990 and again in 2003 would still preoccupy America nearly twenty years later, with little possibility of success.

By the mid-2000s, some members of the Air Force were growing concerned about China's military modernization. General Howie Chandler was among them. He was a product of the post-Vietnam Air Force. He graduated from the Air Force Academy in 1974, joined the flying side of the Air Force, and was trained as a fighter pilot. That meant he flew F-15s and F-16s, the most dominant fighter aircraft of the era. Chandler secured all the flying and staff assignments associated with a first-rate officer, sometimes known as a fast-burner. He completed several assignments in Europe, but most of his operational assignments were in the Pacific. By 2007, Chandler was a four-star general and commander of Pacific Air Forces, the top Air Force position in the Pacific region.

Chandler's career was heavily influenced by an intense Air Force focus on operational excellence. During the difficult Vietnam years, the Air Force suffered significant losses at the hands of the Soviet-equipped forces of North Vietnam. Just as important, the Air Force was frustrated by its inability to destroy fixed targets during the war. Numerous bridges and other fixed targets were left undamaged because the aircraft of the time were incapable of delivering bombs accurately. For that reason, additional missions had to be flown, which exposed more and more pilots to enemy ground fire.

As the Vietnam War drew to a close, the Air Force was determined to correct the problem by improving the weapons, targeting, and, perhaps most importantly, training—the keys to successfully destroying a target. Also, the aircraft had to be better protected from enemy ground fire. The shorthand for this era's vast array of improvements is *precision* and *stealth*. The focus on precision included sending GPS satellites into space. The focus on stealth allowed aircraft to fly more safely in the presence of enemy ground fire because they were not easily detected by radar—a forty-eight-thousand-pound plane appeared like a 1.5-ounce bird on a radar image. These changes—what some would call a revolution in military affairs and others called "the second offset"[14]—were on full display during the 1991 Gulf War. The Air Force suffered losses during the Gulf War, to be sure, including pilots who were taken as prisoners of war, but far fewer losses than ever before experienced in combat. Moreover, modern precision came into full view. Targets were typically destroyed with a single sortie. Far fewer weapons were needed, and the effects on enemy capabilities and morale were devastating. Instead of counting the number of sorties required to destroy a single target, the Air Force and Navy started counting the number of targets that could be destroyed with a single sortie. Airpower had come into its own and became the weapon of choice for US policymakers.

After the Gulf War, the Air Force was called upon in the Balkans, Afghanistan, and again in Iraq. This prompted an intense debate among the military services about the modern use of airpower. Some suggested the Air Force could play the dominant role in future military operations, and others were more cautious in not wanting to oversell the value of airpower. At the extreme, some suggested that armies would be less relevant in the future, which of course was not well received by

[14] The first offset was the reliance on nuclear weapons to offset Soviet numerical superiority.

the Army's supporters. In actuality, modern airpower created a new American way of war. It became the military instrument of choice after it proved its effectiveness in the First Gulf War.

This was the state of the modern US military—the United States maintaining dominance, or what some called over-match, over its regional foes—when Chandler took command of Pacific Air Forces in 2007. But Chandler and one of his key deputies, Colonel Marty Neubauer, knew the Air Force had a problem. They wanted to quietly call attention to the problem before it was too late to fix it.

Chandler and Neubauer were well aware of the earlier Net Assessment war-gaming work that showed China gaining an edge on US military capabilities. Both had held staff positions in the Pentagon when "transformation" became an important part of the DOD lexicon, something of a buzzword if not serious exploration, and airpower had its proponents in the Pentagon, White House, and Congress. They also were in place when their military service had stunning successes in Afghanistan in 2001 and Iraq in 2003 only to then see the successes become stalemates as insurgent forces took root in both war zones. At the time, all eyes were on turning the war around in Iraq with the surge of forces under the command of General David Petraeus. But Chandler and Neubauer's problem was an entirely different matter.

The type of air warfare that produced such huge successes against Iraq in 1991 and later in the Balkans, Afghanistan, and Iraq again involved flying large numbers of aircraft relatively short distances against mostly undefended targets. This was made possible using stealth aircraft very early in the operation to attack known enemy air defense systems, then to follow with large numbers of fighter and attack aircraft and bombers to overwhelm ground targets, including moving targets like tanks and trucks. The advent of precision weapons made this targeting possible and forced ground forces to hide or be subject to withering attack. It was the pairing of modern airpower with modern ground forces that allowed US ground forces to close on Baghdad within a few weeks of initiating the attack.

All this, however, required forward bases for ground and air forces and all the necessary supporting infrastructure—fuel and maintenance, to say nothing of food and water, along with people, equipment, and facilities. At the time of the Second Gulf War (i.e., the Iraq War), US forces could operate with utter industrial efficiency. But forces required a sanctuary—an area from which to operate that was mostly, if not completely, free from external attack—near the area where the fighting was to take place so that US forces could position themselves and be supported as they entered combat. In other words, the sanctuary had to be sufficiently close to where the operations were to take place, but sufficiently far so as not to be subject to disabling attack. Falling in on an infrastructure that had been built over the prior decades by the United States and its Gulf partners, US forces showed once again their prowess in carrying out an attack against a lesser military foe. It was a bit like Mohammed Ali sparring with a less capable boxer. There was little initial doubt about the outcome.

But the possibility of war in the Pacific, or to be more accurate, the possibility of deterring war in the Pacific was a different matter. For years, military planners largely dismissed the idea of a war with China in the Pacific. Although China fielded an enormous army, its air and naval capabilities were vastly inferior to those of the United States and its closest allies. That was the situation China faced during the earlier Taiwan crisis: The United States Navy was too much for the China's military, so China's military backed off. But things were changing. US policy was still focused on doing business with China, but China's military was improving, and significantly so. Lest anyone doubt China's seriousness, it was building the military wherewithal to bring Taiwan to heel. China would not be embarrassed again in a crisis over Taiwan.

Chinese leaders decided, accurately, that the way to keep the United States away from a fight over Taiwan was to threaten US and allied airbases—especially the floating airbases called aircraft carriers, the US Navy's crown jewels. US airbases in the Pacific cover vast expanses of ocean, from Hawaii to Guam to Okinawa in Japan. In the Western Pacific, there is a lot more water than land for runways. The idea that the United States could position its forces close to potential battle areas was out of the question. Whereas the Middle East had a large number of bases that could house military aircraft, the Western Pacific had relatively few. And the airfields that did exist were long distances apart. Planning for operations with the kind of industrial efficiency that was possible in the Middle East would not be likely or even possible in the Western Pacific region. Aircraft carriers could help solve the problem, but they, too, had to steam thousands of miles to be on station and could be at sea only for limited periods of time. Plus, aircraft carriers are vulnerable to missile attack in the same way airfields are, and they are not capable of sustained operations over days and weeks. Munitions run scarce, and planes need fuel and repair. And the United States has only eleven carriers; that number is unlikely to grow even as modern versions come out of the shipyards to replace the aging fleet.

What made matters worse was growing awareness of China's military modernization plans. US military planners had long assumed China's military would undertake efforts to deny US military forces access along China's periphery—an area that is sometimes known as the first and second island chain, with the first island chain encompassing parts of Japan, Taiwan, and the Philippines, and the second island chain extending into the Mariana Islands, including Guam. They could do this by targeting the runways and infrastructure that housed the US arsenal. By the early 2000s, there was growing evidence that China had built sensors and missile forces capable of inflicting severe damage on the first island chain, including Taiwan. There was also evidence that China would soon have the capability of reaching the second island chain.

It was in this context that Chandler and Neubauer began asking serious questions about the effectiveness of US forces in the Western Pacific if they were directed to operate there. They raised these questions at the same time that ongoing operations in Iraq were going poorly, when US and Iraqi casualties were mounting, and it was evident that the United States was playing referee in what had become an Iraqi civil war. In Washington, DC, and especially in the Pentagon, there was little appetite for new problems when the big problem called Iraq was far from under control. As expressed by John Culver, who was the top CIA analyst on China at the time, "Don't tell me about an enemy I'll have in ten years when I'm fighting two wars today."[15]

---

[15] As captured in "China Expert John Culver on Beijing's Military Prowess," *Intelligence Matters* podcast with Michael Morell.

As cover for the new analysis the team at Pacific Air Forces was undertaking, Neubauer would send messages under the subject line of "World Hunger." What Neubauer and Chandler were looking for was what Chandler sometimes called "ground truth" on the state of play they were facing in the Pacific.

One of the challenges, of course, was thinking about tomorrow while dealing with the challenges of today. Chandler recalls being "eaten alive" by "daily ankle-biting things with little time to focus on important tasks." He made clear to Neubauer that it was Neubauer's job to keep Chandler focused on today and tomorrow.

Neubauer had developed close working relationships with the analysts at RAND and the Center for Strategic and Budgetary Assessments, often known as CSBA. What Neubauer wanted to know was whether RAND and CSBA could join forces to develop and run a war game that pitted a modern Chinese force against US forces in the Pacific region. Marshall's earlier games had been speculative, more akin to "what ifs." Chandler and Neubauer wanted to see where things stood against China's modernizing military.

RAND and CSBA teamed up and ran a game over several days in 2008 under the banner of Pacific Vision. In classic war-gaming style, "red" and "blue" teams were presented with scenario details and were instructed to prepare "moves" or responses as the scenario unfolded. A team of specialists assembled to adjudicate the moves and serve as referees to determine outcomes as each team decided how to employ its available military forces. The details of the actual game remain classified, but what was evident was that the "blue" team, composed of Chandler's military planners, was constantly frustrated—an understatement—as it attempted to operate against a modern

Chinese military. Even with some opportunities for "do-overs," the situation was grim. In Marty Neubauer's recollection, "It was really ugly, really fast. We could not go where we wanted to go. There was a lot of blood."

Neubauer went on: "There were shocked people there.... The point of the exercise was to confront people with uncomfortable capabilities." The participants realized that "we don't have anything to bring to the table.... The first hot wash [the military's term for an initial report] was very quiet."

The participants came away realizing their task was not impossible but that changes would be needed if they were to expect success in the future. Over the course of the game, which amounted to several days of frustration, the realization that new thinking and new capabilities were required sunk in. Some of the changes would entail new investment in technology and capabilities—aircraft, missiles, sensors, and missile defenses. Others were more routine but nevertheless essential, such as additional concrete runways, aircraft shelters to protect planes from missile attacks, runway repair kits to get runways back in operation after a missile salvo, underground fuel storage, and so forth. Even such mundane activities as spreading aircraft out across the airfield rather than parking them wingtip to wingtip could make an important difference.

Chandler knew there were some things he could do right away. "There was a certain amount of funding that needed to go to Guam in terms of some [aircraft shelter] hardening. I picked up some 'budget dust' to start on dispersal.... It doesn't take a lot of money, it does take planning and forethought."

What was most important, however, was the realization that the expectations and lessons of air operations in the Persian Gulf would fail against a more capable foe. It led Alan

Vick, one of the RAND analysts who participated in the game, to later conclude that this represented an end to the American way of war.[16] If airpower had emerged as a dominant approach for the use of US military forces, it was no longer going to work.

Or, as John Culver said, "They basically took our way of war and tried to turn it against us. And I think it's kind of an open knowledge now that they [China] have largely succeeded; that a US decision to intervene in a military conflict with China would be one that would be fraught for a US president to decide in a way it wasn't in the mid-nineties."[17]

President Barack Obama had campaigned on the idea of ending the wars in the Middle East. He was looking for an opportunity to show new directions in US foreign policy and chose November 2011 to unveil a new policy in Australia that would soon be known as the "pivot to the Pacific."[18] In the course of the speech, Obama signaled the end to the wars in the Middle East:

> In just a few weeks, after nearly nine years, the last American troops will leave Iraq and our war there will be over. In Afghanistan, we've begun a transition—a responsible transition—so Afghans can take responsi-

---

[16] Alan J. Vick, *Air Base Attacks and Defensive Counters: Historical Lessons and Future Challenges* (Santa Monica, CA: RAND Corporation, 2015), https://www.rand.org/pubs/research_reports/RR968.html.

[17] As captured in "China Expert John Culver on Beijing's Military Prowess," *Intelligence Matters* podcast with Michael Morell.

[18] Barack Obama, "Remarks by President Obama to the Australian Parliament" (press release), White House, November 17, 2011, https://obamawhitehouse.archives.gov/the-press-office/2011/11/17/remarks-president-obama-australian-parliament.

bility for their future and so coalition forces can begin to draw down.

This represented a promise Obama was not able to keep. US forces remained in Iraq and Afghanistan for the rest of his presidency. Still, he was able to confirm his intention to commit the United States to a major change in policy. He went on to say:

> As we consider the future of our armed forces, we've begun a review that will identify our most important strategic interests and guide our defense priorities and spending over the coming decade. So here is what this region must know. As we end today's wars, I have directed my national security team to make our presence and mission in the Asia Pacific a top priority. As a result, reductions in U.S. defense spending will not—I repeat, will not—come at the expense of the Asia Pacific.
>
> My guidance is clear. As we plan and budget for the future, we will allocate the resources necessary to maintain our strong military presence in this region. We will preserve our unique ability to project power and deter threats to peace. We will keep our commitments, including our treaty obligations to allies like Australia. And we will constantly strengthen our capabilities to meet the needs of the 21st century. Our enduring interests in the region demand our enduring presence in the region. The United States is a Pacific power, and we are here to stay.

Obama was announcing a change in policy many had hoped to see, probably none more than Kurt Campbell, who was serving as assistant secretary of state for East Asia. Campbell had trained as a Soviet analyst in the waning days of the Cold War, but then specialized in East Asia as the Cold War drew to a close. He was deputy assistant secretary of defense for Asia and the Pacific at the time of the Taiwan Strait Crisis. He knew the embarrassment China suffered and was more aware than most of China's determination to never let it happen again. Hillary Clinton recruited him to serve in the State Department, and he made China his focus, if not his preoccupation. Nicknames abound in Washington, DC. China watchers are often known as "panda huggers." In the Obama administration, Campbell was known as a "panda mugger."

Campbell was a driving force behind Obama's pivot. It had many dimensions—the remaking of alliances, the building of economic and commercial ties, the cementing of cultural relationships, and importantly a shift in overall national security planning. This is often talked about as a "whole of government" approach, not just shoring up a military deficiency but also strengthening the larger system that underwrites it. Obama introduced a larger foreign policy initiative. The military would play a role, but not the leading or even primary role.

What has become clear over time is that partnerships built around only military ties are fragile at best. Partnerships that are built around larger economic, commercial, and cultural ties are more enduring. NATO offers a good example of strong, interrelated ties, underwritten by a military commitment. Obama's intention in 2011 was to bring a similar focus to the Pacific.

Not everything was smooth leading up to the announcement of the pivot. Although some on Obama's team saw China as a new competitor, a new rival to the United States, others on Obama's team were seeking China's assistance in support of other policy goals. Barry Pavel was a member of Obama's National Security Council staff. He recalls that in the first two years of the Obama administration there were tensions within the NSC staff. The Chinese military had taken actions that the defense team on the NSC staff thought merited protest. China's military actions in the Western Pacific were becoming more aggressive. They were interfering with ship movements and commercial fishing in nearby waters. But, of course, the NSC is made up of different offices with different roles. The China team was focused on smoothing and sustaining relationships. They were less interested in highlighting problems. The defense team was observing a series of provocations and wanted to express concerns through official channels. To be silent, they thought, was to acquiesce.

But the China team on the NSC had different ideas. They were seeking China's help with Iran and North Korea and worried that raising concerns about military activities could interfere with China's cooperation on other matters. This is known as linkage in the world of diplomats. In an exercise of seeking and offering favors, there was little appetite for raising concerns over what were seen as lesser military provocations. There were bigger matters at play.

As Pavel remembers it, "The defense community wanted to raise the provocations with them and push back. And the China people in the NSC did not want to do that." Pavel said he persisted and kept getting the same answer: "We can't do that." So he raised the matter again but was told "if we raise this issue, then it's going to upset the PLA [People's Liberation Army]. And if we upset the PLA, that's going to upset the

Chinese leadership. And if we upset the Chinese leadership, we won't be able to secure their cooperation on Iran and North Korea. So, we can't raise this with the Chinese."

Pavel was stunned, frustrated, even furious, and feeling a bit ornery. So, he said this to his counterpart on the China desk: "Well, if that's the case, why don't we just invite the PLA to the deputies committee meetings and eliminate the middleman."

When Obama announced the pivot, he did not anticipate that he was about to hit a fiscal buzzsaw. A geopolitical one, too.

The fiscal buzzsaw came as Obama battled with Congress over his new health care policy. He was forced to trade spending cuts for an increase to the debt ceiling. What followed from Congress was a sequester that amounted to a set of planned spending cuts that would be rolled out over the next ten years, but with sizable cuts to defense and nondefense programs in the first two years. There was an $85 billion cut in the first year, and $109 billion cut in the second. These cuts might sound small in the context of overall federal spending, but they are enormously difficult to achieve when implemented with little or no notice. Obama's defense team had to make payroll and cover the costs of the ongoing wars. There was not much left for new good ideas.

The geopolitical buzzsaw came from the Middle East. Obama delivered on the troop reductions he promised in Australia. From a peak of 170,000 American troops in Iraq in 2007, Obama returned all fighting forces home by the end of 2011. But after ISIS overran western Iraq in 2014 and was on the march toward Baghdad, he committed additional advisers and ultimately returned upward of 5,000 troops. Even if the American troop presence in Iraq was small compared to the surge of forces in 2007, it still consumed the top national security team that had to deal with the crisis that Obama thought was behind him. The idea behind the pivot was to turn America's vast reservoir of talent and energy to a new set of challenges. If Asia was the future for the United States because of its vast economic potential both as producer and consumer of goods, and the pivot was Obama's answer to the future, a lack of funding and, perhaps more importantly, a lack of leadership time meant Obama would find it difficult to make progress on this important new initiative.

Even still, new problems with China arose, and the Obama team was not always ready. The emergence of Xi Jinping came to be seen as a turning point.

Evan Medeiros was trained as a Chinese linguist and then he immersed himself in China's politics and military developments as a RAND analyst. This work brought Medeiros into contact with most of the China hands in and around government in the early 2000s. It also led him on numerous visits to China to see firsthand how China was emerging as an economic and military powerhouse. When Barack Obama was elected president in 2008, Medeiros joined the Obama administration as a top China adviser on the National Security Council staff. Medeiros's job was to serve as part of the connective tissue between the warning and action machines.

Medeiros was expected to work not just the broad strokes of policy but also the narrow ones, not just the sweeping statements but also the day-to-day actions that formed the basis of policy. Medeiros was part of the action that led to Obama's 2011 announcement of a pivot to the Pacific. He's quick to note that work on the pivot was well underway before Obama's

2011 speech in Australia and smarts at the suggestion that the pivot originated at the State Department. His job was to help make the machinery run, and it was not helpful to have one agency claiming credit for a broad, government-wide initiative.

Medeiros was a longtime consumer of intelligence on China's military developments. During his time at RAND, he was frequently involved in work documenting China's growing military prowess. He was aware of the broad trends and always on the hunt for new information. As Medeiros reflected on his tenure on the NSC staff, he pointed to a time when there was a shift in mood in the United States and a gap in the steady stream of information the warning machine was producing.

One of the issues Medeiros highlighted in his reflections was the striking gap between the national security sector and the commercial community in their attitudes toward China. US national security planners had been worrying about the path China was following since the mid-1980s, when Andy Marshall had formulated his initial analyses. But for a long time, the US commercial sector saw China as a place to do business and it defended its strong ties with China and wanted to benefit from them.

As Medeiros saw it, things changed in 2014. Xi Jinping came into power exhibiting a more authoritarian bent and, more importantly, gesturing toward what some suspected but hadn't yet assimilated. Xi's slogan "Made in China 2025" was a signal to the US business community that China intended to dominate and that the market in China would not be nearly as welcoming to outside business as it had been in the recent past. Medeiros noted this began under Hu Jintao, Xi's predecessor, but intensified with Xi. It was a wake-up call, he said, and led a number of US business interests to conclude that China was a bad place to do business as well as eroded broader business support for the US–China relationship.

To the extent there was a constituency standing in the way of being more vocal about China's military modernization efforts, that constituency chose to mute its voice. In Medeiros's assessment, "The Chinese screwed it up." This took a variety of forms, but theft of intellectual property may have been the biggest. The business community was willing to tolerate intellectual property theft to a certain extent but became much less tolerant when it understood Xi's goal was for most goods to be "made in China," even if that meant China succeeding with other people's inventions. Whereas the business community had been a voice for restraint in US policy circles, tamping down calls for a more assertive military posture in the region, that voice finally quieted.

Tom Fingar made the very same point to us. Under President George Bush (43), Fingar was the first deputy director of national intelligence for analysis and director of the National Intelligence Council, the body responsible for coordinating inputs across the government warning machine. We spoke with Fingar long after he'd left government and posed the same question we posed to Medeiros.[19] When did the business community change on China? Fingar was very clear in his response. Up until the time Xi Jinping took the reins in China, the US business community embraced the idea that China was open for business. More and more manufacturing was taking place in China, not just inexpensive consumer goods but more sophisticated consumer technology. Apple products "designed in California and assembled in China" were commonplace.

---

[19] Tom Fingar, interview with authors, Stanford, June 2019.

The business community understood China's military was modernizing, but it saw it through the lens of China becoming a more normal nation, a "responsible stakeholder," as famously uttered by Robert Zoellick in 2005 when he was serving as deputy secretary of state.[20] Zoellick was arguing that China would need to adhere to the rules if it wanted to continue benefiting from the larger trading system, but reduced to its essence, he was signaling this was a moment to make money, not war. Fingar told us the American business community embraced the idea and wanted China to see the wisdom in what Zoellick and others were offering. "By the time Xi began to put his stamp on Chinese behavior, the business community had tired of Beijing's repeated failures to honor its commitments ('promise fatigue') and incessant demands for access to proprietary technologies and other concessions."[21]

In our conversation with him, Medeiros pointed us to another problem where he believed Obama could have made good on his new policy, but the policy community did not see what was coming. In 2013 and 2014, reports began flowing of China's land reclamation projects in the South China Sea.[22] Dredging equipment was operating along the reefs in the South China Sea turning sunken reefs into habitable islands, islands that could house radar equipment and helicopter landing pads.

---

[20] Robert Zoellick, "Robert Zoellick's Responsible Stakeholder Speech," National Committee on US China Relations, https://www.ncuscr.org/content/robert-zoellicks-responsible-stakeholder-speech.

[21] Tom Fingar, interview with authors, June 2019.

[22] Additional insights in this section come from the Congressional Research Report: *China's Land Reclamation in the South China Sea* (Ben Dolven, Jennifer K. Elsea, Susan V. Lawrence, Ronald O'Rourke, and Ian E. Rinehart, CRS Report 7-5700, Washington, DC: Congressional Research Service, June 18, 2015), https://sgp.fas.org/crs/row/R44072.pdf.

Names like Fiery Cross, Subi, and Mischief were entering the policy lexicon. Over a period of fewer than two hundred days, China's Sky King, a new self-propelled dredger and something of a technological marvel, moved some thirteen million tons of sand and seawater onto the reef, which was more than three times the number of cubic yards of concrete needed to build the Hoover Dam. The problem was that with little or no warning or assessment, the US policy community was left to sit and watch China create new island territories in contested parts of the South China Sea. The area fell under various claims not only by China but also Vietnam, Indonesia, Malaysia, and the Philippines. This island building could have been thought of as a land grab, where China first had to create the land before it could grab it. And it did so with speed, and without apology. By creating new land for runways and helicopter landing pads, it helped expand its defensive periphery well beyond its immediate shoreline. China also made clear to neighboring countries that the South China Sea was more of a nearby lake than a neighboring sea. Anything moving through the South China Sea would be monitored closely.

The US Navy's Pacific Fleet responded by conducting freedom of navigation operations, FONOPS in military parlance, as it does to this day, deliberately steaming ships through contested waterways. This is not something a ship captain does on his or her own. It requires approval all the way to the White House. In this case, moving ships through the contested waterways was more to thumb our nose than anything else. There was little the Navy or other US military services could do to walk back the new fortifications China had built in plain view. Medeiros said this was a moment he would have liked to have called a "mulligan." In Medeiros's judgment, this was the moment when Obama could have made clear the seriousness of the pivot. Even if the United States could not stop China from dredging sand and rock from the seabed, it could have taken other steps, like helping Taiwan harden itself against possible future attack. But the warning machine wasn't clear about what was transpiring, and the action machine was not prepared to take decisive steps by calling China's bluff or even orchestrating a countermove, such as selling more controversial weapons to Taiwan. So, instead, they watched. We could still sense his frustration when we heard Medeiros recount the story some years later.

Economic ties were a centerpiece of Obama's plans, and the Trans-Pacific Partnership, or TPP, was Obama's bold new initiative. It involved remaking trade relationships in the Pacific region with the United States at the center. China could join the agreement but only if it agreed to the terms. The United States and its Pacific partners, including Japan, South Korea, and Australia, had no intention of accommodating China to bring it into the fold. The deal was struck, but in the end Obama could not deliver. Even Hillary Clinton, who was Obama's secretary of state and who was running for president, declared she was not in support of the TPP. In a way, the situation resembled that of Woodrow Wilson and the League of Nations. Just as Wilson could not deliver the League of Nations from the government he led, so, too, Obama could not deliver the TPP, which arguably was the heart of the pivot, from the government he led. Donald Trump abandoned the TPP soon after he entered office.

And how does former Pacific Air Forces commander General Howie Chandler recall Obama's effort to pivot to the Pacific? "We did the worst thing we could do, which was nothing—after we declared a pivot to the Pacific." America lost time—was there any way to get it back?

Sitting in an office building in easy eyeshot if not earshot of the Pentagon, Robert Work, former deputy secretary of defense to Barack Obama and also a retired Marine, known to many as Bob, spoke to us in blunt terms. He sounded more like a field Marine than the former no. 2 at the Pentagon.

"We've been in a strategic cul-de-sac," he said, referring to the nation's twenty-year fixation on counterterrorism following the attacks of September 11.[23] "And it's time we find our way out."

We didn't expect Bob Work to sound this frustrated. After all, this is a man who spent his full military career in the Marine Corps and served all eight years of the Obama administration, first as undersecretary of the Navy and then as the Pentagon's no. 2 civilian leader. Known as a straight talker with a knack for turning big ideas into action, Work's big idea during the Obama administration came to be known as the Third Offset.

Work told us he learned of the "offset" concept from Paul Kaminski, a former undersecretary of defense who was a pioneer during the early days of stealth airplanes and the Global Positioning System (GPS), the same GPS that runs Google Maps and can tell you where to find the nearest Starbucks in less than a blink of an eye, though back then it was still a national security secret. Kaminski explained to Work that stealth and GPS provided an advantage that offset Soviet advances in the

---

[23] Robert Work (former deputy secretary of defense), interview with authors, August 2019.

1970s. What Work was trying to accomplish as deputy secretary was to introduce a third offset. The first offset was the idea that nuclear weapons could offset the massive number of Soviet troops that threatened Western Europe. The second offset was the pairing of stealth and precision-guidance weapons that proved so effective in the First Gulf War, allowing a single warplane to inflict the damage that previously took a dozen or more. This, too, was intended to offset Soviet numerical superiority in Europe.

Work spent his time in the Pentagon making the case that the United States needed to leverage its investments, talent, and creativity in machine learning and artificial intelligence in order to stay ahead of strategic competitors. The United States needed to be a global leader in the area of "human–machine interface." That was what he meant by "Third Offset."

At the same Reagan forum where Mattis spoke, Bob Work explained the difference between offsets and tit-for-tat approaches. He said, "The United States never tried to match a Great Power tank for tank, ship for ship, airplane for airplane, or person for person." The better approach is to identify an offset, an ability to inject a capability or approach that changes the nature of the competition, that puts distance between us and a competitor and keeps it as long as possible. Work was trying to put teeth into the Obama administration's pivot to the Pacific.

When we sat down with Work in 2019, he began our conversation by reflecting on this nation's nearly thirty-year military involvement in the Middle East, with deployments that began soon after the Iraqi invasion of Kuwait in August 1990 and continue to present. This endless and frustrating involvement has cost the United States time, resources, and lives, and

Work bemoaned how little we had to show for the investment. The wars in Afghanistan and Iraq had gone badly, lives were lost, and trillions of dollars had been spent. The Middle East region had siphoned attention from the strategic community—made up of analysts and military planners—over the last thirty years. Three decades of experience and know-how associated with fighting and policing in the desert would offer little of value in a long-term competition with China or Russia. What is worse, according to Work, is that the Chinese and Russians knew this and were eager to take advantage of our preoccupation to get a head start on modernizing their own military capabilities. Work was not so much faulting previous presidents, including President Obama, whom he served, as venting his frustration at how little the United States had to show for concentrating on the Middle East for so long.

Bob Work was on a mission as deputy defense secretary. He brought very specific ideas, namely, that the focus in the next decade needed to be on human–machine collaboration and combat teaming. Work wanted to bring artificial intelligence and machine learning to the modern battlefield. In our meeting, Work cited a now well-known instance of two amateur chess players using personal computers to outplay a field of supercomputers and grand masters separately. Humans and machines together could achieve something that neither could achieve alone.

"The way we'll go after human–machine collaboration is allowing a machine to help humans make better decisions faster." He went on to argue that machines themselves would play more prominent roles, too—reducing risk to humans. "I'm telling you right now," he said, "ten years from now if the first person through the breach isn't a fricking robot, shame on us."

What Work suggested was revolutionary in many military circles. Machines had played important roles in military thinking since the dawn of the industrial era but always at the service of humans. Never the other way around. Bob Work was challenging the Pentagon to think in different and inventive ways. Not everyone in the Pentagon was ready to line up behind him.

Bob Work was following in the footsteps of two former bosses, Andy Marshall and Richard Danzig. Both Marshall and Danzig had been seized with the idea of dissuading future competitors. They were always searching for ways to capture strategic advantage. Work, serving as deputy secretary nearly fifteen years later, knew that if the United States dominated the competition in how humans and machines joined forces, it could yield strategic advantage, or conversely, how failing to do so could create enormous vulnerability. In the 1980s, Soviet leaders had feared that the Strategic Defense Initiative would protect the United States from nuclear attack while leaving the Soviet Union vulnerable. In Work's judgment, the application of artificial intelligence to contemporary military problems might yield a similar kind of advantage. He didn't know the answer. He was instead focused on exploring the question.

In this regard, the growing tension over Taiwan—whether or not Taiwan continues to move toward independence or gets absorbed by mainland China—takes on much more strategic significance than control over a nearby island. Although considerable attention has been given to China's recent military developments, there is also a quickly growing awareness that China presents a competitive threat across a host of political, economic, and social dimensions. How the nation emerges in the aftermath of the COVID-19 pandemic is but one illustration. How the world adapts to the introduction of 5G communications is another. Who gains the upper hand as artificial intelligence and machine learning become more real than imagined is still another. The list of illustrations could go on and on.

Andy Marshall would have appreciated the point. Although he often spoke in the abstract, he reflected deeply on the idea that military competition takes place in the context of the larger political and economic wherewithal of the competing powers. Economies have lifelines—goods, services, raw materials, and trading partners—and lifelines have vulnerabilities. In the mid-1980s, Andy Marshall and his collaborator Charles Wolf recognized the Soviet economy was overstretched and could run out of steam; they saw China as the longer-term challenge. They would now recognize that "Made in China" represents a much broader challenge. Being capable of fighting battles along the periphery of its territory is one thing; controlling elements of crucial global supply chains is another—the latter being far more important over time.

As the dominant producer of semiconductor chips, Taiwan Semiconductor Manufacturing Company—or TSMC, as it is known—represents a critical supply node in the global information economy. TSMC is responsible for more than 55 percent of the total semiconductor market. Because COVID-19 had the effect of initially slowing auto sales, the semiconductor industry shifted production to consumer goods, leaving the auto industry without needed chips when it swung back into production.[24] As of late 2022, the auto industry had yet

[24] "COVID-19's Impact on the Semiconductor Industry," Kidder Mathews, August 11, 2021, https://kidder.com/trend-articles/covid-19s-impact-on-the-semiconductor-industry/.

to recover because of a lack of chips that serve as the backbone of modern automobile electronics systems. This factor has been cited repeatedly for the lack of new car inventories in the United States.[25]

TSMC's production has since rebounded, but the chip shortage has caused some careful watchers to wonder what would happen if TSMC's production suddenly ceased in a fight over Taiwan or if distribution was limited if China took control of Taiwan the way it did with Hong Kong. What effect would that have on the global economy and America's ability to keep its economy churning?

Michael Leiter is among the people posing this question. Leiter is a former director of the National Counterterrorism Center, the government organization created in the moment of finger-pointing after the attacks of 9/11 as the interagency hub for all things terrorism. He is a product of the 9/11 era. He was trained as an attorney and clerked for Justice Stephen Breyer. He was also a Navy officer and known and admired as a skilled operator inside the warning machine. Leiter has a youthful look and a very quick mind. He spent several important years of his life hunting for Osama bin Laden. He's now seized by the idea of maintaining a viable security posture in the Western Pacific because one of perhaps the most crucial elements of the global supply chain sits within a relatively easy bull's-eye of China's rapidly expanding missile threat.

Leiter described a war game he had been asked to participate in. It wasn't a classic war game in the sense of an invading

---

[25] Dylan Walsh, "How Auto Companies Are Adapting to the Global Chip Shortage," MIT Management Sloan School, June 21, 2022, https://mitsloan.mit.edu/ideas-made-to-matter/how-auto-companies-are-adapting-to-global-chip-shortage.

army crashing across the border to lay claim to nearby territory. Rather, it was a game focused on the loss of a critical component to the global economy. Not oil. Not natural gas. Not rare earth materials. Instead, it was microchips. He described the scenario: "It was a tidal wave...a tsunami that kind of wipes out half of Taiwan....They didn't want to do the conflict thing. They kind of wanted to narrowly focus on the loss of the semiconductor capacity and see how the US could react in the world."[26]

In classic Leiter style, he cut to the chase. "Shut down TSMC, shut down the US economy." He went on to describe what happens when production at TSMC ends, be it from an attack or a natural disaster: "The first two years, all right, you're sort of kind of begging and pleading and trying to cajole [in search of needed microelectronics], but you can't actually do much, you're really at a loss as a country for three years plus."

Leiter then said, "In fairness to the US government writ large, I think there has actually been an appreciation...of what the problem, what the vulnerability, is." Congress has become alert to the enormity of the challenge. "So, you've actually... had an enormous amount of, I think, identification of the problem, concerted effort, and some actually bipartisan effort and bicameral effort and cross executive branch congressional effort to try to build the capability. So the good news is, in seven years, we'll be in a much better place." The August 2022 CHIPS and Science Act should give a boost to the domestic semiconductor industry.

Leiter summed his views this way: "You know what the bad news is? The US government only reacts to crises. It just

---

[26] Michael Leiter, interview with authors, November 2021.

doesn't do anything long term anymore. It's got to get either punched in the face or see the guy cocking his arm back to do anything. 9/11 is the perfect example. It wasn't as if everybody didn't know something was coming. Same thing with the pandemic. I mean, you have people to warn you. But until you have the crisis, where's the motivation?"

The implications, of course, are extraordinary.

The United States engaged in the Cold War with the Soviet Union for more than forty years. It was a whole-of-nation effort, not simply involving the US government and military but also the entire fabric of US society. Business, labor, media, schools, churches, and civic groups across the entirety of US society engaged in the colossal effort. The nation formed alliances and trading blocs. Businesses thrived across continents. Europe and Japan bought American products. Americans bought products from Europe and Japan. American soldiers lived in Germany, the United Kingdom, Japan, and South Korea. Their children were born there and went to school there. Citizens groups formed cultural ties. Throughout this time, the Soviet Union and its partners sat outside this thriving, prosperous set of global relationships. The Cold War was the subject of constant national debate. Differences over how to approach the contest divided the nation at times, as with the "red baiting" and McCarthyism of the 1950s. But the Cold War unified the nation at other times, as in the triumphant Apollo landing on the moon—national pride surrounding that moment was palpable and lasting. The entire Cold War hung under the shadow of the threat of nuclear war, with extraordinary measures taken on the part of the United States and the Soviet Union to ensure that neither party had the ability to inflict a disarming nuclear attack without being subject to a devastating nuclear response. In the words of Cold War theorists, this was known as "second-strike stability" or sometimes simply as MAD, which stands for mutual assured destruction.

Yet, despite the totality of the contest and the enormity of effort of two major superpowers, there was never a time when the Soviet Union could deliver a devastating blow to the US economy by attacking a single set of production facilities on an island positioned some hundred miles off its border. There was never a time when either party possessed the ability to create a single point of failure of the entire economic system.

It is a point worth repeating. The United States never faced such a single point of vulnerability throughout the entirety of the Cold War. Today, one set of production facilities sits as a bull's-eye within easy range of China's burgeoning missile arsenal.

This is where the worlds of business and national security collide. As long as the United States and China were engaged in a cooperative effort to grow the global economy, there was little need to worry about a key global production node sitting on an island one hundred miles off China's coast. But the moment the US–China relationship is viewed through the lens of competition and not cooperation, alarm bells suddenly go off. No chess player would begin the game by placing their king in check. Yet TSMC sits just there.

The implications of Leiter's observation are truly extraordinary, even shocking. And it is a problem, or better yet a puzzle, that was all too familiar to Cold War thinkers like Andy Marshall.

China's growing missile arsenal—with or without a hypersonic missile—makes the possibility of an attack something that can't simply be dismissed as the whimsy of a few overly exuberant military planners. In the old, tortured logic of the

Cold War, a future US president might actually face the wrenching decision of trading Washington or Los Angeles or Chicago or New York for Taipei. Or in clearer terms, a future US president might have to consider using nuclear weapons to protect a critical lifeline to the US economy, which in turn could invite a nuclear response from China. This is a choice no US president has had to face since the end of the Cold War.

And the situation is now more complicated. In the past, China thought it was sufficient to maintain a minimal nuclear deterrent, several hundred nuclear warheads that it would keep to respond were China to be attacked. But as China's ambitions have grown, so has its nuclear arsenal. In taking a page out of Russia's playbook, China is giving itself options to respond with nuclear threats that it did not possess even a few years ago. With an expanding nuclear arsenal, China no longer has to keep its nuclear weapons in reserve as a deterrence against a nuclear attack—to respond if it is attacked with nuclear weapons. If it chooses to do so, China could now brandish its weapons as a way to signal to the United States or a US ally in the Pacific not to act in the first place. Said differently, with just a few hundred nuclear weapons, China had the ability to respond to a nuclear attack. With a few thousand nuclear weapons, China would have the ability to threaten nuclear attack. If war was to break out over Taiwan, the United States would have to seriously consider the risk of a nuclear attack, not just on Los Angeles, as a Chinese general recklessly threatened in 1995 but on US military installations in the Pacific.[27] No one

[27] Joseph Kahn, "Chinese General Threatens Use of A-Bombs if U.S. Intrudes," *New York Times*, July 15, 2005, https://www.nytimes.com/2005/07/15/washington/world/chinese-general-threatens-use-of-abombs-if-us-intrudes.html.

should assume that a war over Taiwan would necessarily be limited to the area immediately surrounding Taiwan.

As Russia's war on Ukraine unfolded in the early months of 2022, even the slight mention by Putin and Sergei Lavrov, the Russian foreign minister, that Russia might resort to the nuclear option over Ukraine sent strategic chills across Western Europe and the United States and no doubt was a significant factor in Biden's decision to pledge no American boots on the ground in Ukraine and to reject calls for the United States to enforce a no-fly zone, which would bring US and Russian combat forces into direct contact. This is a lesson that will not go unlearned in Beijing or Washington. Just as Beijing would want to persuade Washington that it is willing to take greater risks in a war over Taiwan, so, too, would Washington want Beijing to understand it has the ability to defend a partner even in the face of a nuclear threat—with training, weapons, and advice, if not manpower. Of course, real choices depend on the details of a particular crisis, and those details have not yet been sorted out. Therein lies an important element of the new danger.

China's test of a hypersonic missile that could circumnavigate the globe may not have been an actual Sputnik moment—the evidence remains to be sorted. After all, a weapon that performs well in a single test is not necessarily judged to be effective. Weapons are typically tested numerous times before they are added to the arsenal. Still, China's hypersonic weapons program demonstrates that China's leaders want to have trump cards to play should the fate of Taiwan reach the point where force is considered a viable option or, worse, a necessity.

So, when pundits talk about the stakes surrounding the fate of Taiwan, they are talking about much more than the fate of a small island democracy sitting off the coast of an authoritarian giant. Unless and until the global economy has another reliable source of semiconductors, Taiwan will remain a strategic flashpoint of extraordinary import. The warning machine has warned. So, too, has China's military.

To mix a metaphor, the best strategic minds in America want to be sure that China's trump cards don't lead to checkmate.

We spoke with Clint Hinote in December 2021, in the midst of a recent COVID outbreak. Hinote's fourth-floor Pentagon office looks in the direction of the Capitol. We called in remotely because of COVID-related restrictions, but despite the problems with reception, our conversation was like a seminar on strategic planning.

Hinote is a lieutenant general in the US Air Force. He signs his messages "Q"—perhaps a reference to the quartermaster character "Q" in the James Bond stories who always came up with extraordinary inventions to keep Britain safe. His official title is Deputy Chief of Staff for Strategy, Integration, and Requirements. His real job is to make problems for others. Hinote is an F-16 pilot and instructor pilot for the F-117 stealth fighter, the aircraft that first saw action and won acclaim in the opening days of the First Gulf War in 1991. For the last several years, Hinote has led a team of Air Force officers with the goal of solving the problem of projecting airpower in the Pacific that Chandler and Neubauer identified more than a decade ago. Some of the officers working with Hinote have call signs like Vandal, Stryker, and Wiki. We did not have to remind Hinote of the lessons Chandler and Neubauer drew from their war games in Hawaii or the many war games that have been run since. Early in our conversation, Hinote summarized more than a decade's worth of experiences in trying to fight against China's military in the Western Pacific. "Not only were we losing the war games, we were losing the war games faster," he said.[28] Hinote has made it his mission to bring the losing to an end.

He described for us the breakthrough thinking on ways in which old and new technology can be brought together to frustrate any attempts China might make to invade nearby territory, especially Taiwan. He described the need for a truly joint command and control system—how forces are coordinated, not as a system with separate nodes for Army, Navy, Air Force, and Marine Corps forces—that uses a mixture of air and naval forces, including submarines. He is looking for an orchestra, not separate bands—or in this case, military services—playing on their own. Hinote said this: "We have to be able to bring the domains together [into] something that feels much more organic, much more coherent." He is talking about a true joint warfighting plan, not the sum of four military service inputs.

Drones—relatively inexpensive drones—are an important new element that could complicate China's invasion plans. In the experiments that Hinote has conducted, drones play an especially important role. If fielded at low cost and in high numbers, they present China with a dilemma that Hinote and other specialists believe Beijing could not effectively counter. China could ignore the cheap drones and let them operate in a contested battlespace, providing surveillance and targeting solutions for other weapons-delivery platforms—both air and sea—in the vicinity. Or China can engage the relatively inexpensive and numerous

[28] Clint Hinote (lieutenant general in US Air Force), interview with authors, December 2021.

drones with relatively more expensive—and limited—defense systems to try to destroy them.

"You have some [drones] that really aren't very capable, but they're numerous," Hinote said. "But there's some reason in which the opponent has to honor [make a plan against] them. So they're producing some effect in the battlespace that forces the hand of your opponent. They either have to expend very expensive missiles to shoot them down, or they have to suffer the consequences of whatever they're doing. An example might be a swarm of fairly short-range drones that flies over the Taiwan Strait and keeps track of all the ships that are coming over. If that were true, then China would have a really good reason to shoot those down. But their ways of shooting them down might be pretty limited. And it might be very expensive."

That is a problem US military planners know all too well. If it takes expensive weapons to destroy inexpensive targets, you are on the losing end of the cost curve with every target that is exchanged. The Russians learned this lesson after their invasion of Ukraine in the spring of 2022, when shoulder-fired anti-tank and anti-aircraft missiles had an outsized effect in blunting offensives by vastly more expensive armored vehicles, tanks, and helicopters. One example: the Javelin anti-tank systems provided by the United States to Ukraine cost just under $200,000 each, for a launcher and a missile. Each missile reload costs less than $80,000. The Russian tanks destroyed cost in the millions of dollars each. The damage to morale is beyond estimates. Military planners call this a "cost-imposing" strategy.

For decades, American military strategists have worried that US adversaries were far better at imposing costs on us than we were on them. Terrorists flew commercial aircraft into buildings on US territory at almost no cost to them and incalculable costs to the United States. For over two decades, insurgents in Iraq and Afghanistan used improvised roadside bombs to kill and maim US forces. Countering these efforts came at considerable cost. Still, in 2021 in the final days in Afghanistan, thirteen American service members were killed by a car bomb, an extraordinary loss.

But the United States has not always been on the losing side of cost-imposing strategies. When President Jimmy Carter first revealed that the United States had produced a stealth aircraft, he did so knowing the Soviet military would have to respond by spending enormous sums to improve its air defense systems. America's early dominance of stealth technology not only imposed costs on the Soviet Union but also did so in the most favorable way possible. It caused the Soviets to expend substantial sums on defending their territory, resources that could not be devoted to building more offensive capabilities. Recall this was during the time when Soviet military forces were stationed in the former East Germany and many worried about the potential they possessed to attack into the West. Some have argued that Ronald Reagan's unflinching commitment to Star Wars technology helped bankrupt the Soviet Union. The development of stealth technology played its own role.

Thinking about cost-imposing strategies was a favorite pursuit of Andy Marshall when he was alive. It has been Hinote's job ever since he was promoted to brigadier general. That is what we mean when we say his job is to create problems.

A centerpiece of this idea is what the Air Force calls low-cost attritable aircraft technology, or LCAAT. Some just call it "the cat" for short. These are relatively small, potentially expendable drones. The idea is to fly and recover the aircraft over and over again. But, as suggested in the name, "attritable," if the aircraft is lost or destroyed, it is affordable enough to be replaced. In this sense, the drones are considered expendable. LCAAT was developed by the Air Force research laboratory and is now being tested.

Tom Hamilton and David Ochmanek became fascinated with the Air Force's cat. They both worked at RAND and thought to themselves, if you have a cat, why not have kittens, too?[29] They began thinking of ideas for saturating the airspace with lots of drones that were perhaps under control of the cat or operating on their own as kittens. Military planners often talk about the "kill chain" when considering how to engage a particular target. The chain extends from the sensors that identify the target to the weapons used to destroy it, and includes everything in between. But Hamilton and Ochmanek also understand that if you break a link in the chain, you have broken the chain. They wanted to create something much less fragile. Their vision was to employ cats and kittens in ways that created a targeting mesh. Unlike a chain, which becomes useless when a link is broken, they want to create a durable mesh that maintains structural integrity even if important pieces are bent or broken. Their idea is to use cats and kittens, in large numbers, to saturate the airspace. They recognize that not all the cats and kittens will survive, but the mesh they create will maintain its integrity, and, more importantly, it will be costly

---

[29] Cats and kittens are described in Thomas Hamilton and David A. Ochmanek, *Operating Low-Cost, Reusable Unmanned Aerial Vehicles in Contested Environments: Preliminary Evaluation of Operational Concepts* (Santa Monica, CA: RAND Corporation, 2020), https://www.rand.org/pubs/research_reports/RR4407.html.

to disrupt or damage. Hamilton and Ochmanek have been part of the experiments Hinote has been running.

We spoke to Tom Hamilton in his RAND Corporation office in Santa Monica. He was surrounded by books and a whiteboard with notes and equations. That didn't surprise us.

Hamilton earned his PhD at Harvard and has been on faculty at Harvard, Columbia, and Caltech. He looks very much the professor and thinker that he is.

We asked him how an astrophysicist came to be working on military strategy problems at RAND. His answer was interesting. "I worked on the evolution of quasars in the early universe, and you know it doesn't really feel like you're doing anybody any good. I'm there . . . and you hang out with fun people, and it's challenging, but at some level, it's just completely pointless. You might as well be figuring out a better way to solve chess problems. . . . So, I definitely was attracted to the opportunity to work on things that were important to the real world."

For the last twenty-five years, Hamilton has been working on a different kind of chess problem. Since joining RAND, he has been helping the US Air Force figure out how it can remain successful against a rising China juggernaut. Hamilton described some of his early forays into the problems the Chinese military poses in the Western Pacific. He noted that when he first began working on China scenarios, the challenge was one of acting quickly on warning of possible attack and getting the needed forces to the theater on time. We might think of this as just-in-time delivery of military forces. "For a long time, the dominant issue was getting stuff into theater. We always had enough stuff in CONUS [continental United States]. If we could just get our F-22s [stealth fighters] there in time, then things were going to be okay."

But as the Chinese military improved, particularly as its missiles and targeting capabilities improved, the problem became much harder. It wasn't just a matter of sending airplanes and tankers close to China's periphery; it was making sure they could operate once they got there. "I would say in the last ten to fifteen years, the problem changed. Even if we could get the F-22s there, would there be any runways for them to operate from?... You know, if they can blow up runways... and it's not easy to shoot down [Chinese] missiles. That's a problem that's been extensively studied... and we haven't solved that problem."

That's what got Hamilton interested in cats and kittens, the same cats and kittens that were so much on General Hinote's mind. As Hamilton moved from quasars to deterring war in the Pacific, which means persuading China's leaders that they would be engaging in a war they could not win, he had the idea that perhaps we should use iPhones to watch for shipping traffic in the Taiwan Strait. Why shipping traffic? Why iPhones? Because an invasion of Taiwan would have to come via ships—warships—and the best way to know if an invasion is coming is to see the ships on the move. But, of course, China wouldn't want to halt all nearby shipping so it could move its warships across the strait. It would want to flood the strait with lots of ships—fishing vessels, transporters, and anything else that floats—so it would be hard to distinguish the warships from all the other floating vessels. Moving warships in a crowded waterway would be something of a "hiding in plain sight" scheme. And it would be hard, if not impossible, to pick one target from another if you were determined to stop the warships from reaching their destination, the Taiwan coast.

Unless...

Unless you had iPhones, lots of iPhones, strapped to small, inexpensive drones flying over the strait, talking to each other and sending signals back to aircraft, submarines, and ships that carry the weapons needed to destroy the warships. Of course, these wouldn't literally be iPhones strapped to drones. But the sensors Hamilton imagined would use iPhone-like technology on small, inexpensive aircraft that could flood the sky over Taiwan and keep track of everything moving in the area. The drones would be so small and light that they could be sent aloft with bottle rockets. Yes, literally bottle rockets.

Thus was born the idea of kittens. "That's really what we've been thinking about in the cats and the kittens. The cats carry weapons. The kittens are all information, and information is where things have really moved. I did some calculations for a game of what does it take to sink a Chinese ship? And it turns out, we were getting the same answers that Billy Mitchell—one of the early pioneers of aerial bombing—was getting back in 1922. You just need so many pounds of TNT near the engine room. Things haven't changed very much. But the information [on targeting] is just revolutionary. It's just completely different than it's been before. And that's what we're trying to think about, how... the new capabilities that are available can be applied to what's really been a central problem for a long time." In Hamilton's mind, this is really a revolution.

"So, it's easy enough for China to put, you know, three thousand things that float into the Taiwan Strait. They just have a huge fishing fleet, you can put corner reflectors on fishing boats, and they look, you know, they look like big bright radar returns, and you don't know one big bright radar return from another. There are all kinds of things they can do. But at the end of the day, they really have to get people in tanks across the straits if they want to conquer Taiwan. And if we can find out where those people in tanks are, then we can stop them."

How would the kittens actually work...or play? "I think five hundred would be enough to do the job. But let's go and put out three thousand. That puts China in a terrible position. They have to destroy all twenty-eight hundred out of three thousand little kittens before they can launch their invasion. And if they do destroy three thousand kittens, first of all, that's a huge clue [that they are serious about invading]. But then we'll just put up another three thousand kittens, and we've done the math. And if their missiles kill the kittens [one missile per kitten], which they can, but the missiles cost a couple of million dollars each and our kittens cost $500,000 each.... We'll just keep throwing kittens at them until they decide to stop shooting missiles."

In other words, it's a viable cost-imposing strategy. American drones with iPhone sensors cost a whole lot less than the Chinese missiles that would be used to destroy them.

We were curious why the idea has so much potential now and not, say, ten years ago. "We could do it ten years ago," Hamilton explained. "We can do it a lot better now. We've got cell phone technology that is much, much harder to jam. What I think is the most exciting technology is that tiny little chip that Qualcomm builds that can do directional antennas in millimeter wave, which is for a variety of reasons basically impossible to jam. It's also a gold mine for us, because it means that our guys can talk to each other over short distances, but they can't be jammed by anything a hundred kilometers away. This is a relatively new technology. Right now, Qualcomm is manufacturing millions of those things every month; they're showing up by the millions."

The technology behind kittens is one thing. Using them is another. How might kittens be used in an actual crisis? The essence of deterrence during the Cold War was that each side was holding a loaded gun to the other's head. The gun took the form of nuclear-tipped missiles and nuclear-weapons-filled bombers that could be sent aloft at a moment's notice. Each side had enough missiles and bombs to destroy the other if it was attacked first. In other words, each side made certain that if one gun went off, the other gun would go off, too. This is what mutual assured destruction means. Or as Albert Wohlstetter, who worked at RAND decades before Hamilton, put it, the delicate balance of terror.

But as Hamilton discussed with us, iPhones in the sky are something very different from a loaded gun. They are taking pictures and sending images back to command centers. It's more like using closed-circuit TV cameras in a crime-ridden neighborhood. Installing cameras might just stop some of the criminal behavior.

We asked Hamilton if he knew what the Chinese military thought of the idea of cats and kittens. He mentioned there had been articles published in Chinese military journals. "They talk about it, they talk about it in considerable detail. And they are concerned. They don't know what they can do about it."

What might China consider doing in response? "What the Chinese can do is they can put a bunch of kittens over the Taiwan Strait, which means we can't invade China. If we launched an amphibious invasion of China, they would know all about it, they could sink our ships. What it really produces is a world in which it's very hard for anyone to launch a big kinetic operation without everyone knowing about it, or at least everyone who was really interested enough to put a bunch of sensors in the air in the relevant area. If China put a kitten mesh over the Taiwan Straits, it would be very difficult for us to deal with it. Fortunately, our defense planning is not based on launching an amphibious invasion across the Taiwan Strait."

Maybe this moves us a step away from mutual assured destruction and a step closer to mutual assured awareness.

Tom Hamilton knew the Air Force was working on the idea of cats, the LCAAT that General Hinote described to us. Hamilton's idea was that the cat wouldn't be of much use without the kittens. It is an idea that has now gained favor in many parts of the Air Force and is being explored in games and experiments. The question remains whether there is sufficient time to give birth to the cats and kittens that could be the answer to deterring war in the Taiwan Strait.

Cats and kittens are only a part of it. Hinote's experiments contain considerable technical detail, and much of the work is classified. But the thrust of what he has been doing is to develop a suite of capabilities that could conceivably impose more costs on US adversaries than they do on the United States, and in so doing he wants to sow doubt, considerable doubt, in the minds of attacking forces. It is what Ochmanek calls the essence of deterrence: the real prospect of failure. In this regard, Hinote's recent experiments show a lot of promise.

In fact, Hinote's work has captured the attention of Adam Smith, former chairman of the House Armed Services Committee. In offering his own description of recent war games, Chairman Smith had this to say:

> The basic idea was what if China goes after Taiwan, how do we deter them? What happens? It was an

> unsatisfying result basically. We were not able really to protect our systems, we weren't able to achieve what we wanted.
>
> Well, the Air Force got clever about a year ago and said, What if we had a lot of unmanned systems, a lot of drones?... So let's put that out on the battlefield.
>
> The most important part about that is that when they put that war game up and they picked the red team and they picked the blue team and they said let's go, the red team looked at it and said we're not going to do this. We can't win in this environment.
>
> Now of course the leader of the game said that's cute, but we're actually going to do this because that's the whole point. So, they engaged. But that's what we're trying to achieve. We're trying to get China and Russia and North Korea and Iran to look at this and go, yeah, no, we're not going to fight because that's not going to come out well for us.[30]

It's clear Hinote and his colleagues are on to something. Their bigger worry is whether the rest of the Defense Department and the rest of the US government, including Congress, will approach the problem with the urgency they believe is required.

When we talked ideas with Lieutenant General Hinote, he was full of energy and passion. He speaks in full chapters.

---

[30] Adam Smith, "Representative Adam Smith (D), Chairman of the House Armed Services Committee" (transcript), Defense Writers Group, Project for Media & National Security, George Washington School of Media and Public Affairs, June 29, 2021, https://nationalsecuritymedia.gwu.edu/project/representative-adam-smith-d-chairman-of-the-house-armed-services-committee/.

Exciting chapters. When we talked about what it would take to get it done—how to get warning and action aligned—he was more circumspect. He gave us the sense that there was momentum moving in the right direction, but he was not sure change would come fast enough, perhaps not even within his own military service. "I do not yet think there is a true sense of urgency about it.... Because the types of change that we'll have to have to do a good job of deterring China, of keeping some level of order, while China rises, the type of change is going to be pretty radical, certainly from the military point of view, and there is not yet a sense of urgency at all levels to align around that level of change."

Hinote sounds a lot like General Hyten, former vice chairman of the Joint Chiefs of Staff. In offering advice to his successor, Admiral Christopher Grady, Hyten had this to say:

> The first thing I'd focus on is that, although we're making marginal progress, the Department of Defense is still unbelievably bureaucratic and slow. So I would encourage my successor in everything that he touches to focus on speed and reinserting speed back into the processes of the Pentagon. Because this country can move fast. We have proved it time and time again. But right now it's so frustrating because the answer to every question I give is, Okay, we need the following capability. How long is it going to take? And the answer is ten years. Ten to fifteen years. And they go through the reasons why. Because it takes two years to experiment, two years for requirements, two years for the budget and then if we go really fast maybe we have four years

> to initial operational capability and another five years to its full operational capability. Every program you look at you can see it that way.[31]

If Hinote's sense of urgency and impatience sound a little like Bob Work's, there's a reason for it. Hinote was Bob Work's military assistant while Work was deputy secretary of defense in the last two years of the Obama administration. They are very much kindred spirits. In fact, listening to Hinote express frustration about how the wars in the Middle East caused the United States to take its eye off the ball was a little like listening to Bob Work or even Andy Marshall. He understood there was widespread hope that China would change over time, that it would take on the attributes of a responsible stakeholder, which so many people expected. But it is common among military circles to say hope is not a plan.

Hinote, like Bob Work and Andy Marshall, wants a plan. His focus on cats and kittens might just be it, or maybe it is just the next play in what will be a very long-term competition. We won't know for some time whether Clint Hinote and his many colleagues in the Air Force have seized on the right idea. What we do know is this: He wants to bring the losing to an end. Cats and kittens might just be part of the answer.

As Russia was at war in its attempt to impose conquest on Ukraine, China's leader Xi Jinping was adding urgency to his soundings over resolving matters with Taiwan. Soon after

[31] Hyten, "General John E. Hyten" (transcript).

Russia invaded Ukraine, President Biden dispatched several former officials to Taiwan to send messages of reassurance. Admiral Michael Mullen, former chairman of the Joint Chiefs of Staff, led that delegation. The former four-star commander was wearing a blue business suit and dark tie. His face was covered with a Honeywell N95 mask. He and Taiwan's foreign minister Joseph Wu exchanged elbow bumps when the US party arrived. He was there to signal more than pleasantries. He was there to signal America's commitment to Taiwan.

In his public statement, Mullen had this to say: "We come to Taiwan at a very difficult and critical moment in world history. As President Biden has said, democracy is facing sustained and alarming challenges, most recently in Ukraine. Now more than ever, democracy needs champions."[32]

It is one thing for a former Navy admiral to visit Taiwan. It is quite another for the sitting Speaker of the House to visit. The last time a Speaker visited Taiwan was before the 1995–1996 Taiwan Strait Crisis. Mullen's trip sparked a reaction in the Chinese press. When Speaker of the House Nancy Pelosi planned to visit in August 2022, it sparked a major international incident. China publicly protested Pelosi's visit, and Xi appealed directly to Biden to stop Pelosi's visit, perhaps not understanding the president of the United States cannot control the travel of a sitting member of Congress. This, in itself, is a worrisome omen that could lead to miscalculation. Biden and his team tried to discourage Pelosi from visiting Taiwan, but to no avail. Pelosi has been a longtime supporter of Taiwan,

[32] "Remarks by Former Chairman of the Joint Chiefs of Staff Michael Glenn Mullen," American Institute in Taiwan, March 2, 2022, https://www.ait.org.tw/remarks-by-former-chairman-of-the-joint-chiefs-of-staff-michael-glenn-mullen/.

and her plans to visit were known well in advance. So, too, was China's response.

No sooner did Pelosi leave Taiwan than China began a major military exercise to demonstrate both its newfound military prowess and its seriousness about stopping any further drift of Taiwan out of China's orbit. China was signaling its displeasure, and it did so in a demonstrable way. Although Xi reportedly told Biden he was not seeking a crisis over Taiwan, he was more than ready to exercise China's newfound military muscle.[33] In a world of hard power, Xi was showing that China was ready to make good on its threats even as the United States was contemplating how to make good on its commitments.

Meanwhile, there is every reason to believe China is learning its own lessons from Russia's war with Ukraine. The long Russian buildup—at first believed by some overly optimistic observers to be just a pressure campaign—allowed Ukraine months to improve its defense posture. China no doubt will learn the importance of acting in a rapid, decisive manner in any Taiwan scenario. Why give your opponent months to prepare?

The Chinese also are likely to review their command and control system, since Russia's heavily top-down hierarchy failed to delegate decision-making to lower-level officers who could respond immediately to challenges on the ground. So many Russian generals had to take to the field to command actions that

[33] Lingling Wei, "Xi Sought to Send Message to Biden on Taiwan: Now Is No Time for a Crisis," *Wall Street Journal*, August 11, 2022, https://www.wsj.com/articles/xi-sought-to-send-message-to-biden-on-taiwan-now-is-no-time-for-a-crisis-11660240698.

a surprising number were killed even in the opening months of the war. If all decisions have to flow back to a central headquarters, fast-paced wars may not be that fast-paced. Mistakes that are not corrected quickly can spiral into defeat.

Taiwan also has important lessons to cull. Taiwan has lots of money and lots of weapons, but many US analysts believe it has the wrong weapons—too much emphasis on fighter aircraft that won't survive the first battle, too little emphasis on missiles that could destroy China's aircraft and ships. If the United States and Taiwan have learned anything from the war in Ukraine, it should be to pre-position weapons that would provide Taiwan with an asymmetrical advantage over China's strengths. In other words, the United States should be selling or transferring anti-ship weapons, such as the Harpoon, and anti-aircraft radar and missiles to Taiwan to blunt any Chinese offensive. Weapons like these would not threaten China directly, only a China that is trying to invade Taiwan. This provisioning should be done before tensions over Taiwan start to boil because moving those weapons after a crisis has risen would only accelerate the crisis. Matching an updated Taiwan arsenal to Hinote's and Hamilton's sensor mesh might be the making of a winning hand.

As World War I drew to a close, the British cabinet faced a depleted Treasury but a modern and triumphant military. The British military absorbed ghastly losses in the Great War, to be sure, but it emerged victorious and with a modern kit of equipment, including tanks and airplanes. In 1919, the cabinet made a momentous decision. It concluded that the risk of another major European war anytime in the immediate future was low, so it decided to prioritize restoring the nation's finances rather than continuing to modernize its military. The cabinet intended to harvest a peace dividend while it could. Plus, there was more than enough to do in policing its ever-raucous empire.

In making the decision, the cabinet acted on another consideration. This was a time of rapid technological change. The internal combustion engine had replaced steam and draft animal power as a dominant source of machine-driven power, and electricity was becoming common in everyday life. Air travel was introduced during the war. The cabinet concluded that if it did not time its military investments wisely, it could well waste funds by investing in the wrong capabilities at the wrong time. Judging that any major threat stood at more than a decade into the future, it invoked a ten-year rule to guide its military planning. Year after year, the cabinet would assess the prospects of a future threat, and if that threat was judged to be ten or more years into the future, Britain would continue to focus its time and resources on the demands of policing its empire, which, in the aftermath of the Great War, included parts of the Middle East.

In one of the odd twists of history, Winston Churchill, who in 1928 was Chancellor of the Exchequer, put forth the position that each year that the cabinet determined the risks of another European war to be low, the clock on the ten-year rule should be reset. Of course, it was Churchill who later as prime minister would suffer the consequences of a British military unprepared for Adolf Hitler and the industrial might of the German armed forces. As British forces escaped crushing defeat at Dunkirk, and as the Battle for Britain raged, Churchill would have to look back at the die he helped cast two decades earlier.

The United States never invoked a ten-year rule regarding preparations for long-term military competition with China. But there were times when it behaved as if it did. By the early 2000s, the warning machine was sending alerts of China's modernizing military.[34] But the demands of the ongoing wars in Iraq and Afghanistan proved all-consuming, and adding a new, difficult problem to the menu seemed out of the question. This was in addition to the financial crisis and deep divisions over domestic budget priorities, which led to a government shutdown and sequestration.

Just as Great Britain was consumed with imperial policing duties in India, the Far East, and then the Middle East, so, too, was the United States consumed with its own policing duties. President George H. W. Bush decided not to topple Saddam Hussein at the close of the 1991 war to evict Iraq from Kuwait. This meant leaving substantial US forces behind in Saudi Arabia, Kuwait, and Turkey to contain Hussein. Clinton and Bush (43) were left to manage the aftermath. This included wear and tear on US military forces that were not only expected to be ready for operations in a host of potential hotspots but also actively enforcing no-fly zones in northern and southern Iraq. This came with real consequences for both people and equipment. As operations wore on, preparing for the future was left...to the future.

In their own way, and for very different reasons, Bush, Clinton, and Bush adopted their own versions of the ten-year rule. Obama and Trump would reluctantly continue the tradition until Biden brought US involvement in the Middle East partially to an end when he withdrew US forces from Afghanistan

[34] As captured in "China Expert John Culver on Beijing's Military Prowess," *Intelligence Matters* podcast with Michael Morell.

in 2021. Still, small numbers of US forces continue to operate in Iraq to this day. A top defense official in the Trump Pentagon once said, "My job is to focus on China, and all I work on is the Middle East."[35] It is too early to tell whether the Biden Pentagon team can escape this trap, the cul-de-sac that Bob Work so aptly described.

It is also clear time is no longer on our side. There is no denying that China has emerged as a dominant rival. There is also no denying that the United States will need to engage China over the long haul. There is no option for containing a power that is so inextricably enmeshed in the global economy. The question is whether China's continued rise can remain peaceful. To the extent that China's leaders are convinced they face a real prospect of failure should they initiate a war, then the warning and action machines will have done their jobs. To the extent that China's leaders believe they could succeed in using military force over Taiwan or elsewhere, then the warning signals that came from Andy Marshall and many others over nearly four decades will have been wasted.

Even today, despite all the proper soundings, it is not clear there is an appetite to change and adapt in ways that circumstances suggest. Lt. Gen. Clint Hinote is one person who has ideas for how to change. He is not alone. There are pioneers like Hinote in all the military services, and people like Tom Hamilton behind them. It is not yet clear whether their ideas will produce the type of change that is needed, and in time to make a difference.

Time may have been on our side. It no longer is. New dangers mount.

---

[35] Comment from a former defense official, January 9, 2019.

CHAPTER 4

RUSSIA

The Danger Hiding in Plain Sight

It is a tale of two Munichs, a city that for decades has hosted an annual meeting of European and American security officials, first under the Cold War–era name *Wehrkunde*, a typical German compound noun that literally means "defense customer" but that is better translated as "military science." After the fall of the Berlin Wall, the conference was broadened from an exclusive club of NATO military leaders to a place where Western Europe and the recently liberated East were all welcome to meet and plan for a continent whole and free. The paradigm change was made formal when the meeting's name was pacified to the "Munich Security Conference." Presidents and prime ministers, cabinet ministers and legislators, entrepreneurs and think tank experts, journalists and national security nerds from forty-plus nations flock each February to Munich, the capital of Bavaria, a prosperous southern state often described as the Texas of Germany, where the locals do in fact speak with the equivalent of a drawl.

A conference keynote speaker on February 19, 2022, was a former comedian elevated to the global stage as president of Ukraine, a country that for months was increasingly feeling the martial death grip of encircling Russian forces. Volodymyr Zelensky appeared wearing a dark suit and dark tie against a crisp white shirt, in clean-shaven contrast to the man who would become a household name and a recognizable, grizzled face set above army fatigues as he Zoomed into world capitals and the United Nations after the illegal invasion ordered by President Vladimir V. Putin of Russia.

"The architecture of world security is fragile and needs to be updated," Zelensky told his audience in Munich, prescient to say the least, coming less than a week before Russian forces crossed into the sovereign territory of his country, an illegal offensive that Western military and intelligence officials said was, initially at least, intended to accomplish a speedy capture of the capital and annexation of the entire nation.[1] "The rules that the world agreed on decades ago no longer work. They do not keep up with new threats. They are not effective for overcoming them. This is a cough syrup when you need a coronavirus vaccine. The security system is slow. It crashes again."

As a result, he added, "We have crimes of some and indifference of others. Indifference that makes you an accomplice. It is symbolic that I am talking about this right here. It was here 15 years ago that Russia announced its intention to challenge global security. What did the world say? Appeasement."

Zelensky had exactly pinpointed the day when the world should have begun taking seriously Putin's intention to exact revenge on the United States and NATO for what he saw as disregarding, even belittling, his beloved Mother Russia: it was, again in Munich, on February 12, 2007.

---

[1] "Zelensky's Full Speech at Munich Security Conference," *Kyiv Independent*, February 19, 2022, https://kyivindependent.com/national/zelenskys-full-speech-at-munich-security-conference.

In 2007 in Munich, all eyes that day were on the US defense secretary, Robert M. Gates, who had recently been installed at the Pentagon by George W. Bush with the mission to save the war in Iraq and fix the one in Afghanistan—and restore relations with European allies that had been frayed, if not shredded, by administration policy, including Donald H. Rumsfeld's dismissal of concerns expressed by longtime allies—he derided them as "Old Europe"—over the invasion of Iraq.

But Gates's arrival was immediately upstaged by the head of the Russian delegation—Putin himself.

Putin mounted the stage in the ornate ballroom of the Four Seasons Hotel and accused the United States of global overreach, of provoking a new nuclear arms race, of destabilizing the Middle East. He recited a litany of US sins (offenses that he would repeat again and again over the following decade plus), including the American invasion of Iraq without permission from the UN Security Council, "almost uncontained hyperuse" of military force, failure to advance arms control agreements sought by Russia, and, especially, the further expansion of NATO to Russia's very doorstep.

"NATO expansion has nothing to do with modernization of the alliance," Putin declared. "And we have the right to ask, against whom is this expansion intended?"[2]

Putin is a native of Leningrad, again called Saint Petersburg, a city victimized by the long and vicious Nazi siege. He reminded his Munich audience that NATO had promised Moscow that NATO expansion—which frightened many average

---

[2] Thom Shanker and Mark Landler, "U.S. Undermines Global Security, Putin Declares—International Herald Tribune," *New York Times*, February 10, 2007, https://www.nytimes.com/2007/02/10/news/10iht-web.0210.security.4545958.html.

Russians, too—would in no way threaten Russian security. Whereas Moscow had kept its promise to remove its troops from now-independent Georgia and Moldova, Putin said, the "so-called flexible, front-line bases" NATO operated contained up to five thousand American troops stationed along Russia's very borders. Whereas Moscow had meticulously complied with all arms control agreements and sought to expand them, Putin claimed, Washington was planning to undermine nuclear stability by deploying anti-ballistic missile defenses in Europe. Whereas the rubble from the Berlin Wall had long ago been hauled away as souvenirs to countries that praise openness and personal freedom, Putin warned, "now there are attempts to impose new dividing lines and rules, maybe virtual, but still dividing our mutual continent."[3]

In a harbinger of his subsequent, broader crackdown on a range of Western news platforms and multilateral organizations as agents of foreign influence, Putin asserted that the Organization for Security and Co-operation in Europe, which had been critical of the fairness of a number of elections in the former Soviet space, was being transformed into a "vulgar instrument designed to promote the foreign policy interests of one or a group of countries." He charged the West with funding nongovernmental organizations that interfered in the internal affairs of other nations—presumably Russia, the former Soviet states, and satellites that Moscow still considered to be within its rightful sphere of influence.

Finally, he looked directly at the American delegation of senators and administration officials who sat together, stone-faced, at the front of the ballroom, including Senators John

[3] Shanker and Landler, "U.S. Undermines Global Security."

McCain, Lindsey Graham, and Joe Lieberman. Putin delivered his blunt and final blow. The goal of these Western leaders, Putin said, was a unipolar world: "One single center of power. One single center of force. One single center of decision making. This is the world of one master, one sovereign. It has nothing in common with democracy, of course."

Putin's unexpectedly scathing remarks were high drama on a global stage, and big news. The assembled reporters banged out copy within minutes. The *New York Times* ran it on page one. Russian officials would later recount that Putin had not delivered the more temperate draft of the Munich speech that staff had prepared for him. He had torn it up and rewritten it, entirely, himself. Whether that story is true—and like so many other tales from Putin's inner circle, the truth may never be known—senior US and Russian officials at first debated whether the speech was an unvarnished expression of what was on Putin's mind. Or was it just high drama? After all, immediately after finishing his anti-Western tirade, Putin left the stage, walked to Gates, shook the defense secretary's hand, and invited him to visit Moscow.

After the Putin speech, Gates sat in his upstairs suite at the Munich hotel running through the options: mount a stiff response in case failing to rebut Putin would prove costly? Was the world now seeing the real Putin? Had the former KGB officer revealed his true stripes, not as a champion of Russia joining the club of nations with a free and prosperous democracy but as a new czar who saw the world as a zero-sum game? Where was the Putin who was among the first to offer assistance to the United States after 9/11, including intelligence on and maps of Afghanistan, and who did not block overflight rights granted by Afghanistan's neighbors, formerly Soviet republics? This

Putin described a new Kremlin line that any advance by the West would require Russia to retreat. And this speech was Putin's declaration that Russia would no longer retreat. Was there a way to show the United States was not keeping Russia down—but would keep Putin in line?

Gates, himself a former CIA director and longtime Russia observer, entered a hushed hall the next morning as government leaders, legislators, and military officers reassembled amid news media speculation on Putin's intentions. Gates mounted the podium to deliver the official response of the United States.

He chose words of velvet, not of steel.

"As an old Cold Warrior, one of yesterday's speeches almost filled me with nostalgia for a less complex time," Gates said. "Almost. One Cold War was quite enough."[4]

First, there was polite applause. Then it escalated to a crescendo.

The Munich moment was seen as a diplomatic triumph for Gates and the Bush administration, which in that moment had few European admirers. Gates had deftly defused tensions with grace and wit; a serious diplomatic rift, perhaps even a crisis, was averted. And, after all, hadn't Putin also just invited Gates to visit him at the Kremlin?

Fifteen years later, reflecting on that momentous day, whose importance to history has been elevated by Putin's nonstop march toward authoritarianism, Gates told us that the choice of how to reply to Putin was his alone. "I didn't ask anybody back

---

[4] Thom Shanker, "Gates Counters Putin's Words on U.S. Power," *New York Times*, February 12, 2007, https://www.nytimes.com/2007/02/12/world/europe/12gates.html.

in Washington," Gates said.[5] "But a big part of my response was the historical moment and the audience—and the audience was probably the most important. This was my very first appearance as secretary of defense in front of our allies. And given my background and my history as a hawk on the Soviet Union, I thought it was an opportunity to provide some reassurance to them, that they had somebody who could think in a nuanced way about how to deal with a challenge like Putin."

Gates said he also had observed with great interest the European reaction to Putin's diatribes the day before. "Here was an audience of Europeans that fundamentally was sympathetic to Putin when he walked in the door, and knew he was facing a lot of challenges and was prepared to be very sympathetic," Gates said. Their reaction to the Putin speech "was almost uniformly hostile," Gates added. "This was not the reaction that he or I, frankly, expected from the Europeans to that speech. It seemed to me this was an opportunity not to take the bait of his speech, but to reassure the Europeans, that we would think about this in a thoughtful way and be careful in the way we responded."

And, in fact, Gates did start the critical process of repairing historic ties to America's closest European allies that had frayed under Bush and Cheney and Rumsfeld, an effort that continued when he was asked to stay on to serve as defense secretary for a president of a different political party, Barack Obama—the first time that had happened. Obama's successor, Donald Trump, went the other way and doubled down on bombast against the alliance, including threats to pull troops from Europe. Yet nobody could have predicted how essential

---

[5] Robert Gates, interview with authors, August 23, 2022.

healthy relations with NATO members would be when the next president, a Democrat, Joe Biden, sought to weld an iron-clad European front to counter Russian aggression and support Ukraine's fight for survival in 2022. Allies matter.

Looking back on the evolution of Putin, or at least the public understanding of his evolution, highlights challenges facing the warning machine and the action machine. Note that this insight into Putin's thinking at Munich was not an intercepted Kremlin directive decoded for analysis at CIA headquarters in Langley or the National Security Agency at Ft. Meade, Maryland. It was not a Moscow mole disclosing some insider insights about Putin's newly hardening line. No, this was Vladimir Putin redefining Russia as once again separate from and aggrieved by the West. Putin's speech was a public declaration that the Kremlin would no longer acquiesce to US global dominance. It would not take long for the United States and its allies to see Putin making good on his threats, threats that Putin made very much in public.

It was years later that Gates first publicly shared his unvarnished personal view. "Everybody is familiar with President Bush's famous line about that," Gates said. "He looked in Putin's eyes and thought he glimpsed his soul. I came back from my first meeting with Putin and told President Bush that I looked into Putin's eyes and I'd seen a stone cold killer."[6]

Gates saw the true Putin and his ambitions. "He's trying to reestablish the Russian Empire," Gates said, "and particularly the Slavic core of the Russian Empire: Russia, Ukraine, and Belarus."

---

[6] Dr. Robert Gates and Dr. Michael Vickers, discussion of war in Ukraine, March 23, 2022, Oh So Social Conversation Series, OSS Society, Vimeo video, 58:10, https://vimeo.com/691826334/06dd75351e.

One of America's most respected Kremlin watchers was not at Munich for the Putin speech that day in 2007. John F. Tefft is like the Forrest Gump of the Foreign Service, showing up time and time again on the front lines of history: He joined the State Department's Soviet Desk in 1983, some of the darkest days of the Cold War, served as the US ambassador to Lithuania from 2000 to 2003, and was subsequently confirmed as the nation's top diplomat to Ukraine, from 2009 to 2013, and then to Russia, from 2014 to 2017.

As Putin was delivering his manifesto in Munich, Tefft was at the American embassy in Tbilisi, Georgia, where he had been posted as ambassador in 2005. He told us that he heard Putin's warning clearly. Although he could not have predicted the exact next move by the Kremlin leader, Tefft was not shocked when—just eighteen months after Munich—Russian forces invaded the republic where he was posted, Georgia, a former Soviet republic whose pro-West, pro-NATO leaders came to power in a color revolution, this one in Rose.

"What you basically saw, in 2007, in his famous speech at the Munich conference, was Putin saying to the United States, 'You can't have a two-pronged policy,'" Tefft told us.[7] He recalled that in that moment, he heard Putin clearly warn that he intended to reimpose Moscow's sphere of influence, if not formal domination, across formerly Soviet lands. To be sure, Putin was not expressing an interest in re-creating the Soviet Union with responsibilities for the care and feeding of distant, different constituent republics. But he was sketching the map

[7] John Tefft (former ambassador to Lithuania, Georgia, Ukraine, and Russia), interview with authors, July 22, 2021.

of a zone of influence and charting a course to reestablishing Moscow as a global power worthy of respect, even fear.

"Putin was saying, 'You have to respect our desire to dominate and control and maintain our hegemony over these states in the former Soviet Union,'" Tefft added. "Now, Putin wasn't trying to necessarily reconstitute the Soviet Union, but he put down a marker in that speech, which we then saw followed up in 2008—with the invasion of Georgia. And then we've seen everything since then."

Tefft has both the personal gregariousness and professional substance that are the tools of all successful diplomats, and he has the added skill of being a natural storyteller.

"My own view of Putin today is that this is a man of the 1980s," Tefft said. "He is what we've always thought: he's a KGB guy whose mind and instincts were formed back in the period when he served in East Germany. And then, in watching the end of the Soviet Union, which he has regretted, as he said, he is really kind of formulating his views and particularly his grievances toward the United States. What we've seen for the last twenty-odd years of his rule is an attempt to try to, you know—if you'll excuse the expression—'Make Russia Great Again.'"

For an American diplomat whose first days on the Soviet Desk in 1983 witnessed the horrific downing of a South Korean civilian airliner, en route from New York to Seoul, by a Soviet Sukhoi-15 jet interceptor, which killed 269 people, Tefft can be fair and balanced in assessing American actions that, justifiable or not, give Putin and his supporters grievances and reason to feel cornered, under siege even.

It is one thing for Washington to pursue policies that are in this nation's interests and those of our allies. But with the collapse of communism in Eastern Europe, the death of the USSR, and the end of the Cold War, Putin became a sore loser because America's political elite, in their triumphalism, were sore winners. In fact, a strong case can be made that the West did not even win the Cold War, but that Mikhail Gorbachev folded a weak hand. Had any other Communist Party official been atop the Kremlin hierarchy, it might have ended with a bloody, spasmodic explosion, and not the silence of the hammer and sickle being lowered over Red Square on Christmas evening 1991.

Indeed, after the terror attacks of September 11, 2001, Putin gave a strong sense of desiring to help—sharing hard-earned knowledge of Afghan terrain and not standing in the way when the United States negotiated temporary basing rights to stage US invasion forces in former Soviet republics of Central Asia near Afghanistan. But it was not long before the Bush administration moved to abrogate the Anti-Ballistic Missile Treaty—and no political entreaties from the White House could convince Putin that the missile defense system the United States had in mind was designed solely to defend against an attack from Iran. It didn't help that the initial architecture called for basing the new anti-missile system in former Warsaw Pact nations, with interceptors in Poland and the radar system in the Czech Republic. (An important footnote to the debate is that neither of those basing plans came about as the design of the system evolved.)

The US-led invasion of Iraq in 2003 was seen by Putin as evidence of America emerging as the sole hyperpower on the global stage, a sense that no doubt was fueled, as well, by the 1999 NATO bombing of Serbia during the war in Kosovo. Serbia certainly fits into Putin's dream of reuniting a pan-Slavic world. And the 2005 defeat of Putin's choice for leader in Ukraine, Viktor Yanukovych, seemed to have truly rattled the Kremlin boss. Ukraine's Orange Revolution gave Putin ample ammunition to blame Western meddling—if, by meddling, he meant governments and nongovernmental organizations supporting the right of citizens of an independent Ukraine to select their own government.

"Putin was clearly pissed off over the US moves into Iraq," Tefft said. "But the Orange Revolution in Ukraine was the failure, his failure, a personal failure, because he went to Ukraine and stood out on the balcony the night before the first election with Yanukovych, endorsing him and putting his own personal support behind him. That really kind of shocked Putin, because, as we've seen, the guy still thinks Ukraine is part of Russia. Now the problem is, most of the people, the vast majority of people in Ukraine, don't agree with that."

Tefft noted that some seismic geopolitical changes that had little to do with the United States also explain Putin's hostility. Putin was shocked to the core by the Arab Spring and what happened to authoritarian governments in North Africa, especially the fall of Muammar Gaddafi, the Libyan leader and Russian client.

"I was told by a number of Russians that Putin had watched and rewatched and rewatched the final video of the Libyans catching Gaddafi in the drainage pipe that he was hiding in, and then killing him," Tefft said. Although Putin is not Gaddafi and Moscow is not Tripoli, the Russian leader's grim fascination was thought to indicate concerns about his own fate. And as became apparent after Putin ordered the invasion of Ukraine in spring 2022, the Russian leader lived in—and mostly ruled from—isolation at his dacha far from the Kremlin on the Moscow outskirts.

At the height of the Cold War, a young Air Force captain, Phil Breedlove, was assigned to West Germany, where he regularly climbed into the cockpit of his F-16 and took off on missions to deter aggression by a larger Soviet force arrayed just across the Fulda Gap in East Germany. He knew that, if required by an attack from the east, those deterrence sorties would become frontline combat missions to fight back a Warsaw Pact invasion. His call sign at the time was "Breed," but later in life his crews renamed him "Bwana." (Breedlove said the name goes back to when the Army Air Corps flew out of Port Moresby. The fighter wing at the time worked closely with the native people of Papua New Guinea, and the name Bwana means "headhunter" or "boss.")

When he retired from the Air Force in 2016, Philip M. Breedlove wore four stars on his shoulder and answered less often to call signs but to one of the most distinguished titles in the military: SACEUR, Supreme Allied Commander Europe, responsible for all military action across the NATO alliance. General Breedlove was NATO commander in 2014 when Russian forces seized and annexed the Crimea and Donbas regions of Ukraine, with Moscow's troops operating in direct combat roles and through Kremlin sponsorship of "little green men," armed troops who wore no insignia linking them to Russia and who were used to carry out hybrid warfare.

Russian action took the world by surprise, and Breedlove agrees that the United States had simply taken its eye off the

Kremlin after the collapse of communism in Eastern Europe in 1989 and the demise of the Soviet Union in 1991.

"The Wall falls. We decide we're going to try to make a partner out of Russia and Putin," Breedlove recalled for us.[8] "We go through a period of what I now call 'Hugging the Bear.' We're trying to hug the bear, bring the bear into Western values, the Western family, etcetera, etcetera."

Then came the Russian invasion of Georgia in 2008, "and I think we got our very first glimpse of who Putin really was. The bottom line is, Russia used its military to invade and change the internationally recognized borders of another nation," he added. "And relevant now is that, holy crap, we just, we just sort of brushed that off our shoulders when it was over. And we just kind of moved on and got back to business as usual, trying to hug the bear and bringing it into the West again."

Six years later, Russian forces made swift work occupying two large swaths of Ukraine—action that also caught Western policymakers by surprise.

Breedlove said he "went off like a well-hit nine iron in a tile bathroom" at the absence of sufficient warning he received as NATO commander in advance of that first Russian invasion of Ukraine. Breedlove said he demanded to know why the United States had been caught off guard. "How in the hell do we get caught so flat-footed about what just happened?" Breedlove asked. "Crimea. Yeah, Crimea. And then a couple of months later, Donbas."

---

[8] Phil Breedlove (former NATO commander), interview with authors, March 25, 2022.

Breedlove recalled for us his tense briefing with senior members of the intelligence community from two separate agencies that brought into stark relief how significantly the American government had disregarded the Kremlin. Breedlove said he was told that, on the day the Berlin Wall fell in 1989, the United States had between twelve thousand and thirteen thousand government analysts staring right at the Soviet Union. "We had a pretty good view of them at the strategic level. And we had a really good view of them at the operational level. And we had a fairly good view of them at the tactical level, especially around the fringes of our NATO allies," he said.

By the time of Russia's invasion of Ukraine in 2014, that number had dwindled to about a thousand, Breedlove recalled being told.

A retired senior intelligence official who worked budget issues in the early 1990s confirmed those trends. This official recalled how the intelligence community moved to reorient a massive bureaucracy away from the decades-long singular focus on the Soviet Union and to a much more complicated world, one with diverse threats. After the collapse of the USSR, the intelligence community's budget commitment went from about 50 percent focused on the Soviet threat to about 13 percent. And that reduced expenditure was further challenged by the reality that the Soviet Union had divided along the lines of its former constituent republics, meaning fifteen *new* targets for gathering intelligence and assessing risks.[9]

Add to that the practical, personnel/career impacts of the reduced priority. An entire generation of people decided, "I'm not going to make my career in Russia anymore. That's an old

[9] Interview with retired senior intelligence official, Summer 2022.

thing. So, I'll do proliferation or terrorism or China." That left a relatively inexperienced and, to be sure, a much smaller veteran cadre watching the Kremlin. The intelligence community is very responsive to the policy priorities that are stated or that it perceives. This is why Washington has, historically, been so bad at anticipating state failure or crisis in Africa or parts of Latin American or smaller nations in other parts of the world. Limited resources. Not enough assets to be everywhere. Washington policymakers tend not to focus on areas that do not regularly touch US vital interests. So, sometimes, things are missed.

Breedlove acknowledged that the reduced commitment of people and money still allowed Washington "a pretty good view of Russia at the strategic level, because we can't lose sight of the nukes. But we completely lost them at the operational and the tactical level."

So when Russian conventional forces and the little green men of its hybrid army captured two important areas of Ukraine in 2014, on his watch at NATO, "We had almost no understanding of where Russia's operational-level forces were—and I'm talking about where, as in geography, and also where, as in what shape the forces were in."

He blames not only Washington's zoom on the Middle East after 9/11 but also Washington's repeated, if underdeveloped, desire to shift focus toward Asia. Perhaps most of all, the nation's leadership—and, indeed, the general public—desired to "change the narrative" from one of America's commitment to the post-9/11 Forever Wars in Afghanistan and Iraq to one of no more wars.

At the time, the capture of the Crimea and Donbas simply were not viewed as vital, strategic interests to the United

States—and certainly not at a level that warranted risking outright conflict with Putin's Russia.

A large number of destabilizing, violent actions taken by Putin have arrived as a surprise to Washington. That was true of the invasion of Georgia a year plus after the Munich speech. It is particularly true in the case of the invasion of Ukraine in 2014. With more attention paid to the Kremlin, the Russian intervention into the 2016 presidential election might have been spotted sooner and Moscow's hacking of US government and private-sector computer systems in 2020—the worst in history—might have been anticipated and prevented.

But attention was not paid at Munich—because in 2007, America's attention was elsewhere: on two festering wars, in Afghanistan and Iraq, and on the growing political discord at home. President Bush, and then President Obama, both thought Putin's complaints were exaggerated, perhaps even mostly intended for a domestic audience across Russia's eleven time zones.

After all, the United States had no intention of attacking Russia; in fact, Russia mattered less and less. They hoped that Putin would eventually understand that Russian national interests dovetailed with US interests on the most vital issues: fighting Islamist extremism and containing the rise of China. They assumed Putin would conclude that economic interdependence with Europe, if not full geopolitical integration with the West, was the only viable road to Russian prosperity. Russia's economy was the size of Portugal's. The economy of Texas, another petropower, was larger than Russia's, with a fifth of the population. Put another way, Russia might have more land and natural resources than any other nation, but Texans were five times more productive per person.

Unfortunately, pointing out these weaknesses to Putin had the opposite effect of attracting his cooperation. For a man who dismissed the catastrophic deaths of World War I and World War II, including the Holocaust, as less tragic than the demise of the USSR, comments made in the West that Russia was no more than an "Upper Volta with rockets" must have been galling.

How could Russia expect the United States to recognize its right to an empire that had been joyfully dismantled by its former subjects?

But Putin did.

The annals of intelligence are full of analyses of such moments. Tipping points, telling points, the moments when history shifts under one's feet. Why didn't Washington listen?

It might be better to say Washington heard but didn't care. Bush, Obama, and the majority of officials who advised them thought they did not need to care about what they viewed as Russian paranoia and nostalgia for its lost empire. Putin, on the other hand, expected the West to recognize that he had more than tripled Russian GDP and had stood the country back on its feet.

Career Kremlin watchers also make the point that many Russians, and certainly Putin, don't think like the rest of us in the West. Russia is a nation that was shaped by a wholly different set of historical forces that prepared the West for pluralistic democracy and realism; Moscow missed the Renaissance, the Reformation, the Enlightenment, and came late to the Industrial Age and the Information Age.

One of those who views Russia and Russians that way is Ian Brzezinski, who has served as a deputy assistant secretary of defense overseeing Europe and NATO policy, worked foreign policy issues as a senior Capitol Hill staff member, and even spent time as a volunteer adviser to the post-communist government of Ukraine. Although he has carved out a notable career, he also got foreign policy in his DNA as the son of Zbigniew Brzezinski, a Polish immigrant who became one of the most influential foreign policy thinkers of his generation—staunch anti-communist, but also a realist in global relations—serving as national security adviser for President Jimmy Carter.

Ian was raised in the Northern Virginia suburbs of Washington, DC, in a housing development carved from farmland and horse country. Today he lives just a few communities over, but reflecting on his upbringing in the Brzezinski household, he told us a story about his father that probably was not part of any conversation in Jimmy Carter's Situation Room.

"The area used to be huge corn, soybean fields, and that sort of thing," he recalled.[10] "It was all farmland then. Now, it's a technology corridor and data centers. It's a different character from those days."

He paused, and chuckled. "And my dad used to shoot woodchucks from a window of that house in Spring Hill. We didn't encourage it, but he just thought it was what you're supposed to do. So, if you can imagine, the window would open—and then a long gun with a scope on it would fire thirty-ought-six against a poor woodchuck." (Another pause.) "We never really talked about it." (Until now, at least.)

Brzezinski's assessment of failures in warning and action on Russia is that multiple White House administrations missed the signs of Putin's aggression because "we have a tendency to assume that our adversaries and our allies will always make

---

[10] Ian Brzezinski (former US defense official and volunteer adviser to the post-communist government of Ukraine), interview with authors, January 28, 2022.

decisions like the way we make decisions, that they see things the way we would."

He recalled the "many times, when those of us who were warning about a Russian invasion of Georgia, a potential invasion of Crimea, we would be told, 'That's just not rational. It's counterproductive. He wouldn't do that. It doesn't make any sense. You're just a Polish nationalist. Get over it. The Cold War's over.' And I think it's not because Poles and people of my ilk sit around hating Russians. Instead, I think some of us took more time to figure out, What are they saying? What do they want to do? How do they look at things? And, therefore, how are they going to act? And most policymakers tend to go, We wouldn't do that. The bottom-line fault of American policymaking is that we too often assume people think like us and make decisions on the same rationale as we would."

And, as was the case with China, he blamed a refocusing of the warning and action machines away from superpower rivalries after the 9/11 attacks.

"It's been exacerbated by the fact that our expertise in Russia has been decimated," he said. "We don't have as many Russian speakers in the Foreign Service, in the intelligence community. Everyone's gone over to focus on counterterrorism, you know, the Islamic part of the world. We don't have those guys and gals who used to sit there and figure it out. 'Who's standing where on the Kremlin wall? What does it mean?' As a result, our ability to understand how the Kremlin's leadership is looking at things and then, thereby, more effectively communicate that to our decision makers, further undercut our ability to recognize the warnings that to others can be so clear."

The problem is that any number of Putin's aggressive actions—from the soft takeover of Crimea to the invasion of

Ukraine, to the spread of little green men ordered to foment unrest in former Soviet states, to election meddling—would have been spotted by the warning machine during the Cold War years, when American intelligence had its focus on the Kremlin. But intelligence personnel and surveillance systems were refocused on terrorism in the Middle East after 9/11. The refocus away from Moscow started much earlier, after the fall of the Berlin Wall in 1989 and the collapse and dissolution of the Soviet Union in 1991.

One longtime government national security leader told us that he actually raised these concerns at the highest level while serving as a member of the Obama administration's prestigious President's Intelligence Advisory Board (PIAB). The organization's charter is clear in calling on it to "assist the President by providing the President with an independent source of advice on the effectiveness with which the Intelligence Community is meeting the nation's intelligence needs, and the vigor and insight with which the community plans for the future. The Board has access to all information needed to perform its functions and has direct access to the President."[11]

The official, Richard B. Danzig, whose previous positions included serving as Navy secretary, the sea service's senior civilian leader, and as a principal assistant secretary of defense for manpower, reserve affairs, and logistics, said he was deeply concerned that a focus on terrorism risked missing other equally or, perhaps, more dangerous threats.

"When I was on the PIAB, I was one of several people arguing that, in the midst of the terrorism stuff, we were

---

[11] "President's Intelligence Advisory Board and Intelligence Oversight Board: Introduction," White House of Barack Obama, https://obamawhitehouse.archives.gov/administration/eop/piab.

underinvesting in Russia," Danzig said.[12] "We were shutting down assets and capabilities that we had, for reasons of underinvestment."

To be sure, a dedicated, if smaller, core group of Russia watchers remained. And one cause of decreasing investments was not the sole, direct fault of a US government decision. An increasingly harsh series of tit-for-tat expulsions of Russian diplomats in this country and US diplomats there sharply reduced, over a number of years, the number of well-trained Americans who could get a fingertip sense of what really was going on in Moscow.

For example, the revelations that Russia hacked the email accounts of the Democratic National Committee and state election boards, planted fake news on Facebook, and used trolls to amplify racial and political divides came as a shock to many Americans. One of the few who was not shocked by Russian meddling is Michael Vickers. "What they'd done in Europe was well known for decades. Buying politicians. Setting up think tanks. Planting false stories in newspapers and on TV. What we did not foresee was that they would try doing it to us in 2016."[13]

We were sitting with Vickers in a quiet restaurant in Northern Virginia discussing what keeps him up at night.

Most Americans do not even know the final job Vickers held before retiring—undersecretary of defense for intelligence—where he was the Pentagon's point man for the SEAL Team 6 mission that killed Osama bin Laden. He is not in the famous

---

[12] Richard Danzig, interview with authors, March 11, 2022.

[13] Michael Vickers (former undersecretary of defense for intelligence), interview with authors, May 21, 2020.

photo released by the White House of President Obama and his top military and national security aides in the Situation Room. No, Vickers was monitoring radio transmissions and reviewing real-time surveillance video inside the mission's true command center, at CIA headquarters. The Pentagon commands the intelligence community's largest share of the vast federal budget for spying, about $80 billion, and manages the most intelligence employees, about 180,000 people.

For a man who once practiced infiltrating Soviet lines with a backpack-sized nuclear weapon, Vickers has a mellow, professorial demeanor. In addition to having Army Special Forces training, he has studied Spanish, Czech, and Russian. Moving to the CIA, he was the chief strategist for the largest covert action in American history, smuggling arms and money to mujahideen battling Soviet invaders in Afghanistan. Along the way, he earned a doctorate in strategy from Johns Hopkins University. (He laughs about his thousand-page dissertation, saying, "It's a good doorstop.")

Vickers's answers to policy questions are disciplined, cautious, and usually organized in two or three parts. He analyzes events in sections and subsections.

During the Cold War, the US–Soviet rivalry was described as checkers versus chess, Vickers said. The United States played checkers, taking bold, direct steps—either straight ahead or occasionally at angles—while the Soviets played chess, looking many moves ahead with attacks and feints and attacks over the long game. That is, until they moved themselves into checkmate.

The two most salient points regarding Putin today are, first, that he is a former KGB officer and so believes in the "force multiplier" effect of covert intelligence operations across the spectrum of Great Power Competition, and, second, that he holds a black belt in judo. The US-Russian rivalry today is boxing versus martial arts. Putin knows the United States remains the undisputed heavyweight champion of the world—our economy, our military, everything. So Putin is using judo, using our own weight against us, to put us off-balance and even flip us on our back. He turns our strengths—our openness, our freedoms—against us. That's what we saw in the 2016 election hacks, but we saw it too late.

As the United States approached Election Day 2020, the real issue was not what Russia had done to the United States, but what it was continuing to do to the United States and nearly all the countries in Western and Eastern Europe: cyber invasions, electoral disinformation, financial enticement of politicians and friendly NGOs, fomenting distrust in democracy, driving wedges among the NATO allies. Such was Vladimir Putin's revenge on the West.

Did US intelligence analysts warn policymakers? If so, was that warning specific and actionable? Was the warning rejected because acknowledging it would disrupt higher-priority policy goals, such as attempting to improve relations with Moscow or cooperate on such issues as Iran or Syria? Or was it a failure of the American action machine to prepare its own "measures short of war" that might have raised the costs of Putin's strategy and made him more willing to negotiate a diplomatic solution?

This is the crucial issue. The Obama administration heard the warning but failed to act in ways sharp enough and strong enough to deter Russia or prevent an ongoing campaign of disinformation and election meddling.

Missing the warnings on Putin came in two varieties. One was acting in spite of Russia's concerns; the other was persuading Russia it had no concerns.

George W. Bush and his advisers thought they could act despite Russia's larger concerns—over deploying American ballistic missile defenses, over invading Iraq, over enlarging NATO to Russia's borders. Barack Obama and his advisers, in particular, his secretary of state, Hillary Clinton, had a different kind of hubris in judging that they could persuade Putin of what was in his interest—or what the United States thought was in his best interest. Clinton is famous for pressing a failed "reset" with Russia. Obama and Clinton realized they were wrong, of course, not least with the reminder when Clinton lost to Donald Trump, in part out of Russian interference in the election.

Russia has been in perpetual war with us. We just didn't realize it or act like it. Ultimately, Russia had to understand there would be a high cost for flouting international norms.

Why was Russia's election meddling handled in a way that, in hindsight, was far too gentle a response? Officials have since revealed they were concerned about raising doubts about the integrity of US elections. But that has been the exact outcome, with Putin and other authoritarian leaders able to say that the American electoral process is no better than theirs. And Trump has certainly made the case that the Obama White House was silent on the warning, and took little action, because they were certain Hillary Clinton would win. And in almost four years of contradictory statements, Trump repeatedly said he did not trust his own intelligence agencies when they described Russian election meddling, even as he courted Putin and said he was tough on Russia.

"The level of distrust between our nations, really Russia and the West, remains deep and profound, and permeates all facets of the US–Russia relationship," said Peter Zwack, a top

Kremlin watcher for the US military who retired as an Army brigadier general.[14] Zwack has carefully analyzed Putin's less-than-war offensive against the United States and, in particular, the Russian president's ad hominem attack on American officials based in Moscow as a harbinger of what more was to come.

Zwack is a brilliant analyst, enthusiastic in ways that might strike a stranger as bordering on eccentric. Conversations with him leave a listener feeling a bit like Marty McFly after the professor has explained how installing a flux capacitor into a gleaming, stainless steel DeLorean allows time travel. The metaphor is apt, as Vladimir Putin is in the Kremlin's driver seat, steering Russia back to the future with one eye, and sometimes both, glued to the rearview mirror.

From 2012 to 2014, Zwack served as defense attaché, the senior American military official, in Moscow. His thirty-four years as a military intelligence and Eurasian foreign area officer included postings to Germany, South Korea, Kosovo, and Afghanistan. One of his more unusual, and seminal, experiences came when, as a young captain in 1989, he studied Russian language not in Moscow or Leningrad but in the Russian hinterland—making friends among average Russians, people he would visit again and again over the next decade. He is fluent in Russian and speaks German, Italian, and, he says, a little French.

"In the United States, as good as we are, we follow the shiny ball of it," Zwack said. "Other than the nuclear threat, how did we lose sight of the ball with Russia?" After the collapse

---

[14] Peter Zwack (retired Army brigadier general), interview with authors, September 14, 2021.

of the USSR and the early years of Boris Yeltsin's leadership of Russia, "I think there was an overwhelming sense that the Russians were remote and diminished. And, oh, well, 9/11. And then Iraq. And that shifted even more of our attention." Military intelligence billets once designated for Russia experts were given over to counterterrorism and the Middle East, with a focus on Arabic.

But these warnings about Russian hostilities, or at least the level of hostilities, are still not wholly understood among this nation's political elites and decision makers.

"I think the Russians are in various stages of conflict with us across the entire spectrum," Zwack said. "They *are* in conflict, now, with us in the West. We have an on/off switch. War. Peace. We don't know how to define the in-between. Russians are comfortable living in that in-between. And in that world, we are outmatched. We're uncomfortable. Our societies can't handle it: free media. We don't lie, we don't cheat. We don't do all these things. The Russians continue to keep key aspects of their society psychologically in a near-war footing."

His warning for the future?

China, he noted, is the rising, even pacing, threat. But he wishes more attention was paid to Russia, even when Putin is not pressing Europe to the precipice of war.

"Even on a bad day, they are the country that can take us off the planet, and we them, in about twelve hours. Full stop," Zwack said. "They are going to look for every edge in this gray zone, to chop us down, weaken us, subvert us, because they know they can't win or survive on an equal playing field with us."

On Putin's list of top ten grievances against the United States, numbers 1 through 7 are the enlargement of NATO. There may be other sidebar issues and rhetorical flourishes, but to Putin, it's all about NATO, and in particular the Atlantic alliance's expansion—first into former members of the Warsaw Pact in 1999 and then, in 2004, into more of those former Soviet allies and even into former constituent republics of the USSR itself. And, along the way, there has been Atlantic alliance outreach and openings to Georgia and Ukraine. It is impossible to overstate how all that animates, dominates, Putin's actions.

After that, little agreement exists among Kremlin watchers. Some note, correctly, that NATO is not an aggressive alliance and has absolutely no designs on Russian territory. States newly independent after the collapse of communism in Eastern Europe and the subsequent death of the USSR have sovereign rights to choose paths forward, including diplomatic and military alliances. At the same time, did each and every new NATO member actually add to the security of the alliance? Or just bring added risk along with a new star on the NATO flag? And could a more cautious alliance expansion, perhaps in some cases with market ties elevated above military treaties, have prevented a wounded Russian Bear from taking such an aggressive stance?

"I had been in Moscow from '96 to '99, and I saw the reaction of the first round of NATO enlargement," said John Tefft, the retired ambassador. "And I'll tell you a funny story. When I left Moscow in '99, they had a farewell party for me, and one of the very senior Russian Foreign Ministry types comes up to me. And he had three or so vodkas, put his arm on my shoulder and said, 'Well, Tefft, good luck.' They knew, even though it hadn't been announced, but they knew I was going to Lithuania. And they said, 'Well, you know, we've always thought those Balts were different, you know, even in the Soviet period, you know, they were different.' "

But those same Russian officials warned him that Georgia, Stalin's birthplace, and Ukraine—breadbasket of the USSR, the spiritual source of the Eastern Orthodox Church—were and are parts of Russia.

"And I can't tell you how many times I've gone back and thought about that," Tefft said. "I had no idea I would ever become ambassador in Georgia and Ukraine, let alone come back to Russia. But that was it right from the beginning. The foreign policy and security bureaucracy, the KGB, all of these guys, they've come up their whole lives developed as part of this Soviet system. And for them, NATO was the boogeyman. And if I had a buck for every time I sat down with a Russian and said to them, 'Okay, tell me specifically how NATO is a threat to you,' and I never ever got an answer to that.... Because it's in their mind. It's part of their grievance, their victimhood. It's part of their mental construct, and it isn't really rational. It's much more emotional."

It is accepted wisdom that one of Putin's motivating reasons for ordering an unprovoked invasion of Ukraine was to prevent another former Soviet republic drifting further westward and toward NATO membership. Some analysts argue that if Ukraine had been brought into the alliance earlier, Putin would not have invaded. But even Ukraine's wartime leadership acknowledged that it still was far from meeting membership criteria. Perhaps the worst thing for Ukraine's safety was what now is being called "dangling," the diplomatic equivalent of being only partially pregnant. NATO officially dangled the prospect of membership to Ukraine, for maybe someday, and Putin had to prevent someday from arriving. In hindsight, the better option would have been Ukraine in or out.

Eric Edelman, a longtime Kremlin watcher, served Gates as the undersecretary of defense for policy, the de facto no. 3

job at the Pentagon. He also has long professional ties to Dick Cheney. He served as American ambassador to both Finland and Turkey, frontline diplomatic posts for observing and assessing Russia. Edelman said that American policy should have been redesigned not to keep Russia down but to keep Putin in line. And he feels that has not happened, at least not in time.

Edelman likes to quote British historian Ian Kershaw, who noted that history is written backward but is lived forward. Looking not only at the evolution of US policy toward Putin's Russia but more broadly at national security decision-making, Edelman notes that "you know retrospectively how the story comes out. But when you're in the middle of it, and you're doing twenty things at a time, your attention is divided. The intelligence is not only imperfect, it's incomplete. And it's never going to be perfect. You can never wait until you have all the facts, because you're never going to have all the facts."[15]

He insists that the United States did not treat Russia so badly in the 1990s. "I was involved in the NATO enlargement decisions. And I thought we went to enormous pains that we didn't frankly have to go to as we enlarged NATO in '97. And I honestly think that the 2004 enlargement under Bush (43) was actually more provocative than what we did in the nineties." He equally dismisses any of Putin's expressed concerns about a US-led NATO ground invasion of Russia. "That's just not happening. And he knows that. The fact is that at the time of the 2014 Russian intervention in Ukraine, we did not have a single tank, the United States did not have a single tank in Europe. And we had one Brigade Combat Team that was ready

---

[15] Eric Edelman (former undersecretary of defense for policy), interview with authors, October 4, 2021.

and could go, if we had to intervene anywhere in Europe. And that was the state of disrepair of our defenses in Europe. So the idea that this was any kind of threat to Putin, I always thought was ludicrous."

Edelman blames the warning and action machines, first, for not paying close enough attention to Putin's broad strategic drive to reestablish Moscow's hegemony over its near abroad—but, second, for not appreciating Putin's deep-seated antipathy to the leaders of all the former Soviet territories who were embracing the West. And, third, for failing to act appropriately.

"I think they missed the whole plot about how much the Russian bureaucracy, national security bureaucracy, hated the Georgians, because of Shevardnadze," the Soviet foreign minister and ally of Gorbachev who played a significant role in dismantling the Soviet order and who went on to be president of an independent Georgia. "And they hated Saakashvili, because he was part of the Rose Revolution, part of the color revolutions, which Putin was absolutely convinced were aimed at him."

Edelman says Putin believes that the color revolutions that swept Georgia and Ukraine and Kyrgyzstan "are some American plot to get Vladimir Putin out of office. There's not a secret or a mystery. This is all hiding in plain sight."

Putin understands—in ways the White House and Pentagon and State Department do not—that wars no longer begin with the first bullet or bomb. The question was never, Will there be war in Ukraine? Putin was already at his kind of war over Ukraine when the United States and the West finally got around to paying attention in 2022.

And to be sure, some day the world may learn that Putin's massive invasion of Ukraine can be ascribed not only to his personality and grievances but also to previously unknown contributing factors. Long COVID has been suggested. Or cancer. Or the impact of self-imposed isolation over two years of pandemic. Or even Putin's awareness of actuarial tables for Russian males and his need to move on an accelerated timeline to create his legacy for restoring the Kremlin's place on the global stage.

But those things are not *required* to explain his order to invade Ukraine. The preconditions and mindset have been evident for years and years. His aggression has all been in a straight line. In fact, some who are promoting the theory of underlying medical or psychological rationale for the 2022 invasion have a conscious or unconscious hope to absolve themselves of not seeing Putin clearly before.

Throughout, Putin has sought to capitalize on his strengths in, again, playing a weak hand. Proximity and force numbers give Putin escalation dominance, at least that is what he thought early on. When war starts is up to him. To Putin, control of Ukraine is a vital—even existential—Russian national security issue. Less so in the West. And domestic opinion in the West, in particular in the United States, is so polarized that NATO nations' presidents and prime ministers are not acting with full support of their populations.

In particular, Putin's pre-positioning of vast and potent combat forces was visible in media and social media—and for months. Yet it was dismissed for months.

"I think it's more than just coincidence that this invasion of Ukraine was geared up in the lead-up to the thirtieth anniversary of the collapse of the Soviet Union, the greatest catastrophe, in Putin's view, of the twentieth century," said Brzezinski. "He's become increasingly obsessed by Ukraine. He was signaling it every year, from 2008, the invasion of Georgia, to the invasion of Ukraine in 2014. This is part not just of his Soviet identity but also his Russian identity. Right?

"So how could we have missed this?"

Partly, it was willful acceptance of optimistic assumptions, what is known as confirmation bias, or more colloquially, "You see what you want to see." After World War II and through the end of the Cold War, no Great Power had gone to war with another Great Power. Even the bloody proxy conflicts—Korea, Vietnam—did not pit two major powers openly fighting each other for prolonged periods around the globe or along lengthy front lines. And, after the Cold War, during the so-called and brief end of history, there was widespread belief that interdependence would prevent Great Power conflict. Russia seemed to be moving toward following global rules, trying on the mask of democratization after seventy years of communist dictatorship. Even if the mask did not become the face, it was hoped there could be a convergence of vision and even of values.

Although Putin seems not to aspire for military or ideological domination of the world, as did the Kremlin's masters during the communist era, his actions provide evidence of his intention to dominate his near abroad, a reassertion of Russia as the controlling authority in a gigantic zone of influence, from Eastern Europe across the Caucasus Mountains into Central Asia—regardless of the sovereign wishes of citizens and leaders of those former Soviet states. Putin's actions can be seen as deliberately designed to create the environment for a new round of Big Power negotiations—for example, another Yalta to divide anew the world, with him in Stalin's seat—not like an old Communist Party general secretary but as something like "the new czar," a title now popular in Western assessments of the Russian leader.

The Biden administration deserves credit for marshalling intelligence on Putin's military buildup around Ukraine beginning in 2021 and for deliberately sharing what it could on Putin's actions and Putin's intentions. Putin wanted to weaken the Atlantic alliance and divide it—but NATO has never been more united and is even growing in membership under Washington's lobbying since the invasion of Ukraine. Allies are providing weapons and training and are joining a sanctions regime. And though Washington initially may have been slow to move substantial amounts of heavy arms to Ukraine, it was right in announcing no US boots on the ground—which could have ignited that direct Great Power conflict that nobody wants. Billions of dollars in weapons did begin to flow, some of them the most lethal in the US conventional arsenal. And, to be sure, the decision against imposing a no-fly zone over Ukraine also was correct. To anyone who truly understands military operations, a no-fly zone may have a defensive intent, but it has to be imposed with offensive military actions—neutralizing adversary radar and guidance and communications systems and a readiness to shoot down adversary aircraft. All that looks quite a bit like war—well, exactly like war.

The Biden administration's restraint in not rising to Putin's bait on threatening a nuclear response to Western support to Ukraine was especially important. The rule has to be: Respond to what Putin does, not what he says. Putin's threats of resorting to tactical nuclear weapons if the war in Ukraine turns decidedly against Russia has not, at time of printing this volume, been matched by the necessary preparations and actions.

Even so, the American military and Washington policymakers must prepare for a potential sea change in the broader risk of nuclear proliferation. Ukraine willingly gave up its nukes not long after independence from Moscow. Had it remained a nuclear power, would Putin be so aggressive today? How does that inform other nations with nascent nuclear capabilities or development programs—North Korea and Iran—and those whose treasuries clearly could allow them to go nuclear with ease, such as Saudi Arabia?

The nation's senior officer for nuclear weapons, Adm. Charles A. Richard, who sits atop Strategic Command, said in late summer 2022 that the United States was "furiously" rewriting its nuclear deterrence doctrine to account for a world presenting rapidly escalating nuclear risks.

"Our operational deterrence expertise is just not what it was at the end of the Cold War," Richard said.[16] "So we have to reinvigorate this intellectual effort. And we can start by rewriting deterrence theory. I'll tell you, we're furiously doing that out at STRATCOM" (US Strategic Command).

Robert Gates, who led two of what Russians call "the Power Ministries," the CIA and the Defense Department, is a vocal advocate for revitalizing American "soft power," as well, to confront Putin.

During the Cold War, Radio Free Europe mixed jazz and news reports (of the truthful kind) in broadcasts to audiences across the Warsaw Pact nations. Voice of America did the same, broadcasting into the Soviet Union itself, as anyone who lived in Moscow then can recall how countless citizens stretched long strands of thin wire over their curtain rods to pick up VOA

[16] Tara Copp, "US Military 'Furiously' Rewriting Nuclear Deterrence to Address Russia and China, STRATCOM Chief Says," Defense One, August 11, 2022, https://www.defenseone.com/threats/2022/08/us-military-furiously-rewriting-nuclear-deterrence-address-russia-and-china-stratcom-chief-says/375725/.

signals on their radios despite Soviet jamming. The power of American culture and American values—and accurate depictions of news developments in their homeland and around the world—significantly eroded the Kremlin's credibility.

Does the United States have the machine in place today to orchestrate a similar information campaign against Putin after the invasion of Ukraine? "The answer is, 'Absolutely not,'" Gates said.[17] The machinery of the United States information agency was dismantled in 1999, with its functions divided between an undersecretary of state for public diplomacy and a new Broadcasting Board of Governors.

"Nobody gave any thought to creating a new twenty-first century structure to take its place, and the government is uncoordinated and has no strategy, and it's vastly underresourced," Gates added. "The departments all go their own way. There's no one in charge of messaging for the United States. The White House does its thing. State does its thing. CIA does its thing. Defense does its thing. Commerce does its thing."

Although the Biden administration is to be commended for rallying American and European support for Ukraine, there is no information and communications machine in place to puncture Putin's new Iron Curtain on information and get the truth about the Ukraine invasion to the Russian people—to the troops who are fighting, bleeding, and dying and to those being asked to enlist; to the mothers and fathers sending their sons off to a euphemistically titled "Special Military Operation," not even called a war; to wives and girlfriends of those deployed.

Most of the messages on the war's failure that are reaching the Russian public are delivered in zinc coffins, although news

---

[17] Robert Gates, interview with authors, August 23, 2022.

of a mass exodus of men fleeing the country to avoid a new Russian draft has begun to penetrate the broader society.

"It's hard for me to believe we can't figure out a way to break through the firewalls," Gates said.

Historic American missions offer examples—far more covert, and quite successful. An example came in June 1979, at the height of the Cold War, when Pope John Paul II made his first visit back to his native Poland after being chosen to lead the Roman Catholic Church. Moscow opposed the visit, and the communist government in Warsaw kept his itinerary secret from the public in hopes of diminishing crowds of worshippers who might flock to greet their returning favorite son.

The CIA came up with a plan: agency technical officers developed a suitcase-size TV broadcast system that took over the signal of Polish state television for about ten minutes and broadcast the pope's entire itinerary.[18] Millions of people turned out to greet him.

Another central challenge in confronting and containing a rogue nuclear power like Putin's Russia is sustaining public resolve at necessary levels. The streets of Europe and America may be brightened with the blue and yellow flags of Ukraine, but that feel-good symbolism will not stall any Russian offensives. It takes money and weapons and intelligence and a willingness to sacrifice easy access to supplies of oil and gas—and even to take the hits on the global economy. Sustaining the depth of early support for these measures will be the real test for the West.

Putin has seemed intent on carrying on with the war in Ukraine despite the effects of sanctions on his population. The

---

[18] Interview with former officials with firsthand knowledge of the operation, Summer 2022.

so-called oligarchs have no sway. The early street protests in Moscow and Saint Petersburg, which seemed so inspiring on social media in the first days after the invasion, represented a tiny percentage of a decimal point of the overall Russian population. It is possible, though, the mass conscription, which resulted in thousands of military-age men fleeing Russia and which outraged their families, may have some effect on Russian leaders around Putin. Even so, given the realities of centralized, authoritarian rule over the far more fractious Western coalition supporting Ukraine, Putin can still hope to maintain escalation dominance. The conflict ends when he decides to end it, when he has had enough—or is gone.

The race is to see who gets exhausted first—Ukraine, NATO and the West, or Vladimir V. Putin. Either a victorious Putin or a beleaguered Putin is still a dangerous Putin. If he succeeds in Ukraine, his appetite for adventure may grow. If he fails in Ukraine, his temptation to lash out may be uncontainable. No serious policy can overlook the danger that Russia poses to NATO and the larger international system.

PART III

NEW WEAPONS AND NEW THREATS

CHAPTER 5

GERMS

Saving Lives, Avoiding Extinction

The human species, like all species, is in a constant struggle to survive. It has adapted and evolved over time to become resilient against many forms of disease, though not all forms. But resilience takes time, often considerable time. Modern health practices have aided greatly—amazingly—that adaptation in the form of vaccines to prevent disease and pharmaceuticals to aid in the recovery from disease. And humans themselves have exploited disease to dominate others.

Disease has always traveled at the speed of commerce. In medieval times, bubonic plague spread along major trade routes, where rats and fleas accompanied humans and their goods. In the nineteenth century, the steamship is blamed for the survival of an aphid from California that made its way to France and then wiped out ancient vineyards across Europe, a now-forgotten catastrophe known as the Great French Wine Blight. Today, exotic pathogens are whizzing even faster through twenty-first-century global supply chains. They take jets, long-haul trucks, and fast container ships, lurk inside raw materials and unpasteurized food products, smuggle themselves into commodities, or hitchhike on insects and birds. There is more spillover because we are living in closer proximity to nature that we once left alone.

As the national security machine was built, there was always a concern over biological threats. Fort Detrick, Maryland, housed the US Army biological weapons program from 1943 to 1969. The program was established during World War II by President Franklin Roosevelt, who feared German experiments with biological weapons, and was ended in 1969 by President Richard Nixon before the United States entered the Biological Weapons Convention in 1972. Fort Detrick still houses the US Army Medical Research Institute of Infectious Diseases, or USAMRIID, as it is known. It plays a small but critically important role as a sentinel for the most important biological threats. USAMRIID gained notoriety in the 1990s when the heroics of a small team of USAMRIID scientists and physicians were featured in the best-selling book *The Hot Zone*.[1] Their job was to extinguish a potential Ebola outbreak in nearby Reston, Virginia, where laboratory monkeys that had recently been shipped from Africa were dying of the disease. The terrifying story was later made into a television mini-series.

Of course, the larger public health system is built around Health and Human Services, the federal Centers for Disease Control and Prevention (CDC), and the many state and local public health departments. The CDC was established in 1946 in Atlanta to stop the spread of malaria in the United States. Within a short time, the CDC would expand its focus to all communicable diseases. It is the primary federal government interface

[1] Richard Preston, *The Hot Zone* (New York: Anchor Books, 1995).

with state health departments.[2] Since its inception, the CDC has played a crucial role in disease eradication in the United States and abroad. It helped to successfully eliminate malaria, polio, measles, and smallpox, to name just a few successes.

But the CDC was not built as a crisis action machine, which became so apparent throughout the COVID-19 pandemic. The CDC acknowledged its shortcomings in August 2022, when CDC director Dr. Rochelle Walensky announced a series of reforms. In a mea culpa to the agency's employees, Walensky admitted, "To be frank, we are responsible for some pretty dramatic, pretty public mistakes, from testing to data to communications."[3]

Of course, the CDC is not responsible for guiding public health responses. That is left to the state health departments, which operate under different rules and authorities across states, and this can be the source of uncertainty, confusion, and perhaps even panic in times of a true national emergency. The question "who's in charge" is always left hanging in the air when it comes to health-related crises. Despite the various reorganization efforts that are underway, it still has not been sorted out. The federal government plays a critical role in warning of threats and channeling resources during a crisis. The success of Operation Warp Speed, which produced a viable COVID vaccine in record time, would not have been possible without coordinated federal response. But the federal government does not

---

[2] "Our History—Our Story," Centers for Disease Control and Prevention, https://www.cdc.gov/about/history/index.html.

[3] "Walensky, Citing Botched Pandemic Response, Calls for C.D.C. Reorganization," *New York Times*, August 17, 2022, https://www.nytimes.com/2022/08/17/us/politics/cdc-rochelle-walensky-covid.html.

have an operational arm—something akin to the military—to mobilize public health personnel and resources in a crisis. That is the role of the states.

Preparing for pandemic disease is one thing, and it will be years before the lessons of the COVID-19 experience are sorted out and understood. Preparing for biological attack is quite another and has had up-and-down attention over the years.

The First Gulf War in 1990 and 1991 brought renewed attention to the threat of biological warfare and biological terrorism.[4] As US troops amassed along the Iraqi border, attention quickly focused on Iraq's biological weapons programs, particularly Iraq's ability to infect US forces with anthrax through the use of aerosol dispensers. General Norman Schwarzkopf was responsible for the overall operation, but a small section of Defense Secretary Dick Cheney's policy organization focused on preparations for a possible biological attack. This was led by Scooter Libby, who would later go on to serve as Cheney's chief of staff when Cheney became vice president.

Libby's direction to his staff was simple and straightforward: "If we go to war and Saddam uses biological weapons, I want to be able to look into the eyes of the soldier and their families and say: 'We did everything we could.'"[5] This ultimately led to a hurry-up effort to vaccinate US troops deployed to the Gulf region, particularly those in rear areas—the stationary critical logistics hubs so necessary to support the war effort—who were judged to be most vulnerable to a biological attack.

---

[4] Parts of this section draw on the excellent *Germs: Biological Weapons and America's Secret War*, by Judith Miller, Stephen Engelberg, and William Broad (New York: Simon & Schuster, 2001).

[5] Miller, Engelberg, and Broad, *Germs*, 109.

In the end, Iraq did not draw upon its stocks of biological weapons to attack US and coalition forces. Explanations abound about why. The fact that Saddam Hussein's regime was never directly threatened is as good an explanation as any, though some believe the veiled threat of a possible nuclear response, delivered by then Secretary of State James Baker, contributed to Saddam's restraint.

This very particular threat, plus the realization that the recent collapse of the Soviet Union left parts of the vast Soviet biological weapons program unprotected and unemployed, sparked a decade-long effort to shore up America's biological warfare preparedness. The Bush administration was replaced by the Clinton administration in January 1993. They took up the mantle.

In 1995, a Japanese doomsday cult called Aum Shinrikyo carried out deadly sarin attacks in the Tokyo subway. This sounded a different alarm within the national security community on the risks of biological attacks on the civilian population. Then Navy secretary Richard Danzig was one of the people listening. His concern was about overall preparedness. He knew he could not solve the problem on his own, but he concluded doing something was better than doing nothing.

Danzig's style, as we have seen elsewhere, is to raise attention across a wider audience, get other people involved in solving his problem, in putting his ideas to work. In this case, he took a very unorthodox approach.

"I was trying to persuade the Joint Chiefs of Staff [that more attention was needed,]" Danzig told us.[6] "One idea I had in desperation was to get Richard Preston, who was a very

[6] Richard Danzig, interview with authors, March 11, 2022.

good journalist, and who had written this stunning book about Ebola. So, I said to Richard, why don't you write a novel about bioterrorism that can engage people. And he wrote this novel called *The Cobra Event*."

Danzig was looking for something of a calling card. If he could not persuade the Joint Chiefs on his own with intelligence reports and Pentagon briefings, perhaps he could motivate them by bringing in a storyteller, someone whose fictional account brought the threat to life in ways classified intelligence reports could not.

"I gave the novel to not only some members of the Joint Staff who I needed to try and persuade but also to the wife of one of them," Danzig added. "The long and the short of it is this kind of imaginative scheme yielded no consequence in my efforts to persuade Joint Chiefs."

Danzig was willing to accept good fortune over guile. "Completely unexpectedly from my end, Bill Clinton read this book. And Clinton got animated on this.... A request came from the White House to make a video about bioterrorism stuff. So, they produced this little video because Clinton wanted to see it in video form....

"My point is, I think out there, you're like Johnny Appleseed. You throw your seeds out there and where they grow, you don't know. And there's no world in which I thought, I'll try and stimulate Preston. And that'll result in the president doing something. So, the world is just too complicated, too complicated for confident prediction, too complicated for ability to produce these outcomes, in some straight-line way.... You've got to figure out some way to move the system."[7]

---

[7] Richard Danzig, interview with authors, March 11, 2022.

As a result of Danzig's efforts, and that of many others, all US troops would be vaccinated for anthrax. This was a controversial decision at the time. Some serving personnel refused the vaccine, which brought the matter to the courts, but the action was judged on balance an important measure to shore up a particular weakness in US military readiness.

Danzig, of course, was not alone in spearheading a response. William Cohen was secretary of defense during Clinton's second term as president. It was Cohen who famously went on a Sunday morning news program with a five-pound bag of sugar to illustrate the point that small amounts of highly infectious material could have a devastating effect on a concentrated urban population. "Anthrax," he said, stunning the normally loquacious Sam Donaldson and Cokie Roberts into momentary silence. "If Saddam Hussein spread this amount of anthrax over a city the size of say, Washington, DC, it would destroy at least half the population of that city. . . . One breath, and you are likely to face death within five days."[8] Some experts disputed Cohen's depiction, claiming ten times the amount of anthrax would be needed, which is small consolation, but they accepted the urgency Cohen was bringing to the problem.

By the late 1990s, White House staff were simulating exercises involving smallpox and a recombinant virus that would have equally deadly effects. The scenario involved an outbreak in California and along the Mexico–US border. Participants playing the roles of state and local officials were overwhelmed by casualties.[9]

---

[8] Miller, Engelberg, and Broad, *Germs*, 216.

[9] Miller, Engelberg, and Broad, *Germs*, 232–233.

Clinton requested additional funding to improve overall bioterrorism preparedness. At the Pentagon, William Cohen directed the National Guard to develop the capability to respond to bioterrorist threats at home. Still, the vast US public health system was far from ready to meet this growing threat. The information age was penetrating most of society, but it had not yet made its way to the public health system. "Compared with health care in other economically developed states, the U.S. public health system was a disaster. Almost half of all local health departments did not have the use of e-mail; at least one thousand of them had no access to any on-line or Internet service; 20 percent of them still had rotary phones."[10]

"The Centers for Disease Control and Prevention's epidemic surveillance system, created a half century earlier to detect germ warfare attacks, was in tatters. Most epidemiological investigators and even emergency room physicians were unlikely to recognize a case of anthrax, a rare disease. Nor would they correctly diagnose a patient with smallpox, which had theoretically been 'eradicated.' "[11]

One result of this attention was that a national stockpile was established for vaccines and antibiotics. This would be enough to provide forty million doses of smallpox vaccine. For a time, the warning and action machines were awakened and taking reasonable action. The machine was highly fragmented, but it appeared to be moving in the right direction.

In 1998, Dr. Nicole Lurie accepted a position as principal deputy assistant secretary for health at the Department of Health and Human Services. Dr. Lurie is known to her

---

[10] Miller, Engelberg, and Broad, *Germs*, 236.

[11] Miller, Engelberg, and Broad, *Germs*, 237.

colleagues as Nicki. She attended college at the University of Pennsylvania and stayed on at Penn to receive her MD, then went into practice at the University of Minnesota, where she also joined the school of public health.

Nicki Lurie is always on the go, and she manages to go where there are interesting problems to be solved. She went to HHS to work on health disparities among disparate populations. She would find herself working on influenza pandemic planning. She remembers trying to get the machine moving.

"When I was in the Clinton administration, we first started talking about pandemic influenza preparedness, and a huge amount of work has been done," she said.[12] She was among the people who knew it was nearly a century since the H1N1 influenza of 1918, and the world's luck was not going to last forever. There had been major influenza outbreaks since then, to be sure, but nothing on the scale of the 1918 pandemic that caused approximately five hundred million to be infected across the globe and fifty million to die. Some 675,000 Americans died during the pandemic.[13] Lurie knew the country wasn't ready for another pandemic, and she would do what she could from her new position in government. "You know, you can argue about varying degrees of success on surveillance, countermeasure development, stockpiling, deployment of health care, and overall healthcare preparedness. We were trying to get the public health system to function." She was part of the larger

---

[12] Nicole Lurie (former principal deputy assistant secretary for health at Department of Health and Human Services), interview with authors, April 2022.

[13] "1918 Pandemic (H1N1 Virus)," Centers for Disease Control and Prevention, reviewed March 20, 2019, https://www.cdc.gov/flu/pandemic-resources/1918-pandemic-h1n1.html.

effort to bring a slumbering giant awake. But progress wasn't uniform.

In August 1999, several elderly patients were presenting encephalitis-like symptoms in Queens, New York. After several weeks of uncertainty and testing that involved the CDC and USAMRIID at Fort Detrick, the contagion was determined to be West-Nile–like virus. Those involved at the time saw the scramble as a good test case for a potential bioterrorist attack in the United States. The machinery came up wanting.

One particular lesson stood out. "The daily conference calls involving as many as one hundred officials from eighteen different local, state, and federal agencies—often lasting up to two hours—left officials with less time to deal with the crisis. In jurisdictional disputes, it was sometimes unclear which agency in which city, state, or county was in charge."[14]

Twenty years later the same problem would surface again, though at a monstrously large scale as COVID-19 engulfed the planet. Lacking clear lines of authority to act in a crisis means no one is really in charge. Actions are slowed or never taken. People in positions of responsibility are confused. Supplies and materiel do not move to where they are needed. Case numbers grow, and people die. Acting in a crisis through a highly decentralized system is a sure way to produce slow, fragmented results.

The 9/11 attacks changed so much about life in the United States, including a significant renewed attention on biological threats. The Bush (43) national security team was aware that al-Qaeda was experimenting with biological threats, including anthrax and smallpox. A simulation had been undertaken at one of the Washington, DC, think tanks, the Center for

---

[14] Miller, Engelberg, and Broad, *Germs*, 261.

Strategic and International Studies, in collaboration with Johns Hopkins. The exercise was known as "Dark Winter," and it simulated the effect of a smallpox attack in the United States, on Oklahoma City, which up to that time had suffered the deadliest act of homegrown terrorism. Former Senate Armed Services Committee chairman Sam Nunn played the role of president in the simulation, and the results were shared widely within the national security community, including with Vice President Cheney's chief of staff Scooter Libby.

Cheney would later note that among the biggest concerns coming out of the simulation was how many people were vulnerable to the disease because they had not been vaccinated or, for those who had been vaccinated, their immune systems might no longer ward off the disease.[15] This led to a major focus on vaccinating US military personnel. Bush and Cheney themselves decided to be vaccinated as an example to those who were being asked to do the same.

The defense secretary at the time, Donald H. Rumsfeld, was so concerned about this threat that he briefed NATO allies on the biological war game during a stop at alliance headquarters in Brussels. He told the assembled defense ministers that this growing bioterrorism threat would require preemptive strikes against shadowy terror networks and any nations that might supply them with weapons of mass destruction.

Disclosing more details about the bio-war-game scenario, Rumsfeld said that terrorists released smallpox in shopping malls, infecting three thousand people. Within sixty-eight days, three million Americans are infected, and one million die. Vaccine supplies are insufficient. Riots start. The virus spreads to

---

[15] Dick Cheney, *In My Time* (New York: Threshold, 2011), 385.

dozens of other countries. To further make his case, Rumsfeld told NATO defense ministers of top-secret threat assessments by American intelligence. A Pentagon official said those were even more alarming.[16]

With all this as a backdrop, Bush decided to make his case in a public statement in December 2002:

> Since our country was attacked 15 months ago, Americans have been forced to prepare for a variety of threats we hope will never come. . . . One potential danger to America is the use of the smallpox virus as a weapon of terror. . . . Today I am directing additional steps to protect the health of our nation. I'm ordering that the military and other personnel who serve America in high-risk parts of the world receive the smallpox vaccine, men and women who could be on the front lines of a biological attack must be protected. . . . This particular vaccine does involve a small risk of serious health considerations. As Commander-in-Chief, I do not believe I can ask others to accept this risk unless I am willing to do the same. Therefore I will receive the vaccine along with our military.[17]

All this was happening as anthrax was shipped via US mail to government facilities, congressional offices, ABC, NBC, and CBS newsrooms, even the *National Enquirer* in Boca Raton,

---

[16] Thom Shanker, "Reporter's Notebook; On Tour with Rumsfeld, the Jacket Stays on and the Monkeys Stay Away," *New York Times*, June 16, 2002, https://www.nytimes.com/2002/06/16/world/reporter-s-notebook-tour-with-rumsfeld-jacket-stays-monkeys-stay-away.html?searchResultPosition=1.

[17] "President Delivers Remarks on Smallpox" (press release), White House of George W. Bush, December 13, 2002, https://georgewbush-whitehouse.archives.gov/news/releases/2002/12/20021213-7.html.

Florida. The first known exposure was within a month of the 9/11 attacks. At least seventeen people were infected with inhalation anthrax, and at least five people died. Millions of dollars were spent in the cleanup of the facilities where the anthrax was present, including the Brentwood postal facility in Washington, DC. The FBI undertook a multiyear independent review of the case and concluded that Dr. Bruce Ivins, a government scientist at Fort Detrick, Maryland, was solely responsible for the attack. The report was completed in 2010. Ivins committed suicide in 2008.[18]

This prompted Bush to announce in early 2002 that he would be requesting $11 billion to invest in safeguards against bioterrorist attacks.

Throughout all this, a cadre of people was giving constant attention to the growing threat. Colonel Jerry Jaax was the hero of Robert Preston's *The Hot Zone*. He led the team that entered the facility in Reston, Virginia, to take possession of the Ebola-infected monkeys. His wife, Nancy, was part of the team at Fort Detrick who had identified the Ebola virus in the monkeys.

We spoke with Jaax a little more than a year before the outbreak of the COVID-19 pandemic. He was waiting in a booth at a sports bar like any other in a suburb like any other, although this one was outside Kansas City. He had been given the smallpox vaccine since, like so many others, he was in the Army during and after 9/11. But unlike very many others, he also has been inoculated against a full encyclopedia of infectious diseases and has been exposed to several of the worst during times of frightening outbreak. He is among a tiny cohort who has

[18] "Timeline: How the Anthrax Terror Unfolded," NPR, February 15, 2011, https://www.npr.org/2011/02/15/93170200/timeline-how-the-anthrax-terror-unfolded.

served as one of the military's premier commanders focused on readying the force against biological and chemical attack.

Even in retirement, Jaax maintains the cut and bearing of an Army colonel. Nonetheless, it was hard to imagine this soft-spoken, gray-haired grandfather stepping into a chem-bio space suit to go into battle against the outbreak of Ebola among infected monkeys in Virginia, barely a half hour from the White House and just minutes from Dulles airport, one of the busiest international airports in the world.

When we spoke with Jaax, he expressed enormous frustration at how the threat of biological attack by terrorists—let alone the risk of naturally occurring lethal pathogens seeping into the United States—had been all but forgotten, or at best moved to a back-burner concern.

"We started to get ready, and I think we were headed in the right direction," he said, noting how the national vault of smallpox vaccine grew from twelve million to three hundred million by a decade after the attacks of 9/11.[19] But a decade more is almost past. Vaccines expire. And little has been done to condition Americans to accept a controversial vaccination, even if the nation came under attack. Of course, Jaax's warning became all too evident when it came time to roll out the COVID-19 vaccine. The risk associated with COVID-19 is certainly not as great as that of smallpox, but the loss of life has been staggering and the pain across the global economy is undeniable, and resistance to being protected against disease, and to protect others from contracting the disease, is widespread.

"We're just not ready today," he said. "We could be. And we should be."

---

[19] Jerry Jaax (retired Army colonel), interview with authors, January 2019.

When Donald Trump was elected president, the risk of non-traditional attack on the homeland was hardly a priority. The outgoing Obama administration—recalling the calm, professional handover of national security planning from the Bush administration, despite their sharp political differences—planned a similar handoff, including training sessions for the incoming Trump national security team. Lisa Monaco, who served as Obama's homeland security adviser on the National Security Council, held a series of table-top exercises, briefings, and handoff conferences for their incoming Trump administration counterparts.

One, in particular, stands out as worrisomely emblematic to those retiring Obama administration officials. It focused on the threat of global pandemic. Michael Flynn, the incoming national security adviser, spent the session working on his BlackBerry, recalled participants. Other members of Trump's incoming team hardly took notes. They just did not overtly appear interested in the threat of pandemic or other risks not related to traditional state actors, participants said. Little did they know that this would consume the Trump presidency in its last year in office.[20] Nicki Lurie was there. "It was really appalling. And I remember the face of the education secretary getting in the elevator afterwards and talking about what a waste of time this was. It was horrible. We ran them through a big natural disaster and the beginning of a pandemic. They could have cared less."

The US national security establishment recognizes three kinds of severe biological threats. The first is naturally arising

---

[20] Author interviews with participants in the meeting, who spoke on condition of anonymity. Interviews conducted in February, March, and April of 2017. Details of the meeting subsequently leaked, including a March 2020 article in *Politico*, "Before Trump's Inauguration, a Warning: The Worst Influenza Pandemic Since 1918," https://www.politico.com/news/2020/03/16/trump-inauguration-warning-scenario-pandemic-132797.

pandemics—primarily those affecting humans. The second is biological error—that a genetically altered organism could escape from a laboratory or mutate in an unpredictable fashion, wreaking plague on plants, animals, or humans. The third threat is deliberate bioterror. There is a natural human tendency to minimize the seriousness of a threat or minimize the implications. Social scientists term this type of cognitive error "normalcy bias."

The easier and more available gene-editing becomes, the harder it is to predict the likelihood of a bioterrorist attack that could emanate from almost anyone, anywhere. Naturally, experts disagree about how likely the threat is and what should be done about it. Melinda Gates has said publicly that a bioterrorism attack is the threat she worries about most in the next ten years—so much so that she doesn't often talk about it. She and Bill Gates repeatedly warned that an airborne pathogen, whether designed by a terrorist or nature, could kill more than thirty million people in less than a year.

But other experts see nothing in the terrorists' handbook that would indicate a serious effort on attacking the United States with a biological or chemical weapon. "If you're looking at what motivates a terrorist, they want to get a large casualty count. Infection is slow," said Peter Chalk, an expert on counterterrorism and other strategic issues. "It wouldn't be the dramatic explosion that captures headlines immediately."[21] Danzig agreed with this logic.

Yet ample evidence exists that the al-Qaeda cell in Yemen purchased tons of castor beans to experiment with making

---

[21] Chalk, Peter, "Hitting America's Soft Underbelly: The Potential Threat of Deliberate Biological Attacks Against the U.S. Agricultural and Food Industry," RAND Corporation, 2004, https://www.rand.org/pubs/monographs/MG135.html.

ricin, one of the deadliest toxins on earth.[22] And as ISIS was routed from its strongholds in Syria and Iraq, advancing troops gathered proof that the terrorist organization also was experimenting with how to harness naturally occurring viruses and toxins for attacks on Western targets.

This isn't to suggest that nobody is watching.

The Centers for Disease Control are charged with being America's watchdog on the matter. The CDC is a centerpiece of the warning machine. But the CDC has many responsibilities, including monitoring the seasonal outbreak of influenza—no small matter—and other infectious diseases that represent here-and-now problems. It has been on the front lines of COVID-19 for more than two years now, and it has suffered its fair share of criticism. Some argue the CDC has taken on too many disparate missions and should be focused more singularly on "tracking and helping to stop disease."[23] Full stop. Watching for future threats—especially those that may be manufactured by terrorists in far-off locations—is part of the CDC's mission, but the CDC doesn't have its own army, and it relies on the cooperation of foreign partners to gain information about global health threats. Pinning down the origins of the COVID-19 virus shows just how hard that cooperation can be.[24]

---

[22] Eric Schmitt and Thom Shanker, "Qaeda Trying to Harness Toxin for Bombs, US Officials Fear," *New York Times*, August 12, 2011, https://www.nytimes.com/2011/08/13/world/middleeast/13terror.html.

[23] "The Centers for Disease Politics," editorial, *Wall Street Journal*, August 18, 2022, https://www.wsj.com/articles/the-centers-for-disease-politics-rochelle-walensky-center-for-disease-control-and-prevention-covid-vaccines-11660859044.

[24] Amy Maxmen, "Wuhan Market Was Epicentre of Pandemic's Start, Studies Suggest," *Nature* 603 (February 27, 2022): 15–16, https://www.nature.com/articles/d41586-022-00584-8.

In the aftermath of the anthrax attacks, attention focused on creating institutions within the government to help lead a response in the event of a bioterrorist attack. This required legislation and funding. First came the Public Health Security and Bioterrorism Preparedness Response Act of 2002. The act had a number of important features. It established an assistant secretary within the Department of Health and Human Services who would be responsible for coordinating activities in the event of a national emergency. It established a strategic stockpile of vaccines, pharmaceuticals, and medical supplies that would be available in the event of a national emergency. It also created authorities for the federal government to provide assistance to state and local authorities in the event of a national emergency, including the ability to provide funding for local emergency planning and the establishment of the National Disaster Medical System, which allows for medical professional volunteers to surge areas of greatest need in an emergency. This authority, of course, proved invaluable during the COVID-19 pandemic.

The legislation was updated in 2006 with the Pandemic and All-Hazards Preparedness Act. One of the most important features of this legislation was the establishment of the Biomedical Advanced Research and Development Authority, or BARDA as it is known. BARDA is responsible for the development and purchase of vaccines, drugs, medical supplies, and other countermeasures to deal with the effects of a bioterrorist attack as well as pandemic influenza and other emerging diseases. While COVID-19 was not on anyone's mind when BARDA was established, the 2003 outbreak of SARS in China was seen as another wake-up call.

When we asked Richard Danzig for his thoughts about the level of preparedness for the COVID-19 pandemic, he judged the country was much better off because of the earlier actions. No one would say the United States was operating with a well-oiled machine by any stretch, but some of the basic pieces were in place. Operation Warp Speed, for example, which produced the COVID-19 vaccines, including the novel mRNA vaccines, would not have been possible had these earlier institutional pieces—the building blocks of the emergency response machine—not been in place.

It could have been better, but it worked when we needed it to. Had the threat not been taken seriously a decade and a half earlier, the pieces would not have been in place.

Dr. Nicki Lurie returned to government in 2009 as assistant secretary for preparedness and response in HHS. The job is often referred to by its initials—"asper." It was created as part of the post-9/11 reforms, and Lurie's job was to be sure the country was ready for a major health event. Like Danzig, she recalls that many important pieces were being assembled, slowly but surely, over time, including reaching down to states and localities. "Some of my fondest memories being in ASPR were when people stopped me in airports and would say, 'So you're that person who's responsible for making us do those tornado drills or those exercises, and we cursed you. We hated every minute. But you know what, a tornado destroyed our hospital, and we saved lives.' "

Lurie respects the military's focus on readiness. Perhaps some of that comes from the fact that, as a member of the US Public Health Service, which is under the direction of the Surgeon General, she was part of a uniformed service. The US Public Health Service is one of the nation's uniformed services. The tradition of wearing uniforms dates back to the 1870s.

Lurie might not have referenced the lessons the military took from Task Force Smith in Korea, where the units in the initial response were outgunned and outfought, but she knows what it means to be ready, and consistently so. "I think one of the real challenges in this whole area, and especially in the public health and healthcare preparedness, you know, you live through these boom-and-bust cycles of funding, or these cycles of panic and neglect. If you are in a health department, you live off the fumes of the emergency supplemental from the last crisis until the next crisis hits, and six to nine months later you get some influx of funding. Well, you can't play in a system like that."

She knows continuous readiness means not everything will be used, stockpiles will expire, vaccines and treatments will pass their expiration dates. But, again, she compares public health preparedness to military readiness. "Yet people say, 'Oh, my God, we wasted all this money we invested, you know, bought all these monoclonal antibodies. We bought all these vaccines. We did all this stuff. We didn't use it. What a waste.' When we build a jet fighter, and we don't have to deploy it, we don't say it's a waste, right? But we don't necessarily think about preparing for epidemics and pandemics and these kinds of things, as we think about preparing for other national security issues, and all the downstream consequences of these things. And we have a really, really short memory."

Lurie had her own challenges as ASPR, though nothing of the magnitude of the COVID-19 pandemic. She was able to see how the system did and did not work.

Recalling the 2013–2014 Ebola outbreak in West Africa, Lurie observed, "We were sort of watching it. And one of my staff came and said, 'Did you notice how many healthcare workers are dying in Liberia?' And we looked at him and realized this is a really significant problem. We reached out to CDC; they didn't reach out to us. When there was the case in Lagos, people really got spooked. At that point, I said, 'Let's scour the universe and see if anybody's working on an Ebola vaccine.'"

The next step was something Lurie had taken from her long years in the Public Health Service. Get things in motion right away. Time lost likely means lives lost. She likes to talk about on-ramps and off-ramps. She was looking for an on-ramp.

"We had a collaboration between departments of public health and defense departments in four countries where we had this classified database of what everybody was working on. We went in there and, lo and behold, here was this Ebola vaccine candidate made at the Public Health Agency of Canada that had been on ice for ten years. We called them up and said, 'What's the story?' And you know, within about twenty-four hours, I made the decision that we were going to need to get that out of ice and see if we could develop an Ebola vaccine."

But there's a wrinkle to the story. Her authority to act was limited. The legislation following the bioterrorism scares of the 1990s and 2000s gave her the ability to act with speed against bioterrorism threats. Funds were specifically set aside for just these cases, but not others. Not influenza. And certainly not COVID, which no one knew was coming.

"I could only do that because for whatever set of reasons Ebola was classified as a bio threat. So, I could spend bioterrorism money on it." She had an on-ramp and took it. To Lurie, that is the essence of preparedness and response. Whether that on-ramp is the right way for the federal government to function remains an open question.

This is also Lurie's critique of the COVID-19 response. She praised the idea that Operation Warp Speed produced a COVID vaccine in record time. But she thought the effort was too slow to get off the ground. She wanted the Trump administration to act sooner, to take an on-ramp as soon it was clear something big was happening. She was always looking for an on-ramp. You can take the next off-ramp if nothing comes of the warning. "If they had put Warp Speed in place two months earlier, think of how many lives we would have saved. And so the philosophy was, and we did this a lot for influenza, you have to take the on-ramp, you can always take an off-ramp, you can do it really fast. But you can never make up for lost time."

When we returned to the issue of overall public health readiness, Lurie had this to say: "I have argued for a really long time that there just needs to be something like a standing emergency infectious disease fund so you can always take that on-ramp without having to ask anybody. That's pretty straightforward. We need a modern public health system in every sense of the word. It has to function day-to-day, as well as in crisis. It should be able to manage an opioid epidemic . . . it should be able to manage a pandemic. That's a really big piece that we have to put in place. And that is a national security issue. In my mind, it's a national security imperative."

The federal government does not control the state health departments, but it can establish guidelines. That is how Lurie would create more coherence in a national response. She would like to see guidelines flow down to states and localities, not by having the federal government take control, but rather by establishing standards and protocols. Adherence to standards can be linked to the flow of federal funding. "There's no single person or office at HHS that's responsible for public health. It's all over the place. It's completely disconnected from the health-care system, which is crazy. And it doesn't have mechanisms to create expectations, accountability, standardization, at the state and local level. So, you can leave federalism intact. And you can require people to meet core capabilities, you can have an accreditation system, and you can provide a set of incentives for people to achieve those capabilities, which would make it very attractive to them. You can upskill the workforce, there's like a bunch of stuff that you could do. So, you need a modern public health system. It's going to take money and some reorganization."

In July 2022, the Biden administration announced plans to reorganize the Department of Health and Human Services to elevate Nicki Lurie's old job to a new federal agency on par with the Centers for Disease Control and the Food and Drug Administration. The new agency will be known as the Administration for Preparedness and Response, retaining the initials ASPR, which are well-known in the health preparedness community. Not all were thrilled with the change. A *New York Times* article cited a former CDC adviser who noted that "future administrations will rue the day they clipped the CDC's wings."[25] Identifying a lead agency is a logical step. Ensuring the seams that run across agencies are managed in a coherent way is another crucial step. It is not clear the July 2022 plan accomplishes this important step. The fact that other affected agencies may have been surprised by the news suggests the seams have not all been sorted out.[26]

---

[25] "The Biden Administration Is Elevating a Division of H.H.S. to More Broadly Oversee Pandemic Responses," *New York Times*, July 20, 2022, https://www.nytimes.com/2022/07/20/us/politics/hhs-aspr-biden.html.

[26] "Biden Administration Is Elevating a Division of H.H.S."

Moreover, Lurie is quick to point out that even if the United States gets its house in order, it still depends upon the rest of the world for surveillance and warning. It is one thing to have Twitter feeds about people who are sick and dying in far-off places of the world. It is quite another to have the kind of warning that is necessary to get on the on-ramps that Lurie sees as so essential to an effective response. The health system needs to know enough to realize it is looking for an Ebola vaccine that a partner might have on ice somewhere. The Ebola vaccine is not going to help if the real threat is influenza or another contagious disease.

In other words, the machine needs the right machinery. It's not there yet. International relationships will not be built in a crisis. Relationships have to be forged well ahead of time if they are to function when they are needed.

Lurie offered one last piece of wisdom: "One of the most important things you have to do in a crisis like this is maintain the trust of the public. No matter what, no matter how bad things are, you need to maintain the trust of the public. Look at Volodymyr Zelensky, president of Ukraine. Things are horrible. And he's maintaining the trust of the public."

For the warning and action machines to function as they should, they need infrastructure to support them. Yes, infrastructure in the form of bricks and mortar, but also infrastructure in the form of the best human minds available. Just as there was a whole laboratory system to support nuclear weapons development and space exploration, so, too, must there be a laboratory system to support an understanding of bioterrorist threats.

The fear is not just that terrorists might infect thousands by using biological weapons in major cities but also that they could attack the nation's agricultural infrastructure by introducing plant or animal diseases or pests or poison food or water.

In 2004, a Homeland Security Presidential Directive ordered a range of steps to protect agriculture. The bureaucracy continued to chug away at the problem. In 2017, the Office of the Inspector General published an unflattering review about the state of US preparedness to deal with agricultural terrorism. It found that the Department of Homeland Security hadn't done enough to ensure that the US Department of Agriculture (USDA) could "defend the agriculture and food systems against terrorist attacks, major disasters and other emergencies."

Congress would later let DHS, an agency that many believed already had too much on its plate, hand over responsibility for agricultural preparedness to USDA. But the scientific arm of the USDA is considered chronically underfunded and increasingly overwhelmed by the proliferation of plant and animal diseases. Moreover, USDA is simply not a core national security agency, leading some to predict that its issues would inevitably fail to rise to the attention of top policymakers in Washington. Money sometimes does not match mission.

In 2017, Congress also passed a more expansive law, the Securing Our Agriculture and Food Act, which ordered the government to run a robust surveillance system that would provide early detection of diseases of plants, animals, wildlife, and humans. In response to fears from US cattle producers about the rising risks of foot-and-mouth disease (FMD) spreading from China, Congress appropriated $50 million for FMD prevention efforts in the 2018 farm bill.

In 2018, the Trump administration released its first National Biodefense Strategy—a document full of acknowledgments of potential threats to animals, plants, and the food supply, as well as human health, but the strategy lacked an implementation plan. According to RAND researcher Peter Chalk, who testified to Congress about the biological threat back in 2003, fifteen years later, very little action had been taken.

"This stuff is going to bite us," said Ron Trewyn, a former cancer researcher whose life mission became attracting research money, talent, and more public attention to the battle for biosecurity.[27] The mystery was why this was proving so difficult. In the case of African swine fever, it wasn't that the warnings seemed too theoretical or hysterical or that warners were thought to nurture an unpopular political agenda. (These are textbook reasons why bureaucracies have dawdled in the face of impending threats.) And it wasn't that the infection was rampant before authorities even knew what it was. (That happened in the 1980s, when young gay men began dying of AIDS in large numbers a year before the HIV virus was identified.) This time, the warners were well-known scientists publishing meticulous, peer-reviewed work, and their work was read by regulators as well as industry experts. In the specialized lingo of the national security world, credible warning came well to the "left of boom"—meaning that on the timeline of events leading up to a disaster, warning preceded explosion.

"The US is really taking the 'head-in-the-sand' approach," said Trewyn, who was based at Kansas State University.

Yet there is one very tangible symbol that United States government has imagined a darker alternative future and has at last embraced three key conclusions: that plant, animal, and human

---

[27] Ron Trewyn (former cancer researcher), interview with authors, January 2019.

health are interconnected; that all are vulnerable to emerging infectious diseases; and that it will take a concerted federal effort to protect the nation's food supply and environmental health.

From the outside, the symbol doesn't look like much. It is a modern office building rising on a nondescript hill in a not-very-famous town in Kansas, the locals call it the "Little Apple." In fact, the low profile is part of the plan. The $1.25 billion National Bio and Agro-Defense Facility (called the NBAF) is a high-security compound being built by the Department of Homeland Security 120 miles west of Kansas City.

The NBAF is located on the campus of Kansas State University in Manhattan, Kansas. When it opens no sooner than late 2022, NBAF will take the place of the near-decrepit Plum Island Animal Disease Center in New York, which was built in 1954 to do classified research on animal pathogens during the Cold War. The secrecy surrounding the Plum Island facility, as well as its location right off the coast of Long Island, eventually made it unpopular with its neighbors. In Manhattan, Kansas, the facilities are seen as a boost to the university and the farm economy, though some neighbors believe—erroneously—that the lab is working on bioweapons. To be sure, a pathogen that occurs in nature can be weaponized by terrorists as can a pathogen synthesized to do harm, and so the research to defend against these risks overlaps.

The project has not been quick or easy. A congressionally mandated review found a 70 percent chance that foot-and-mouth disease could escape from the laboratory sometime over the next fifty years. That finding prompted a large number of expensive biosafety improvements, which have brought the estimated release risk down to 1 percent over fifty years—but at a cost of more than $800 million and several years' delay.

Some critics think the Plum Island facility could have been rehabilitated for that sum. Others note that there is no federal budget allocation for the extensive maintenance that the NBAF will require, year in and year out.

For now, the NBAF's security is unobtrusive but state of the art. Among many other safety features, the building is engineered to withstand a direct hit by a tornado without releasing a single germ. Inside, the NBAF will have laboratories capable of safely studying the world's most dangerous plant, animal, and insect-borne pathogens—the ones scary enough to be categorized as Biosafety Level 4. It will try to develop vaccines as well as cures for these plagues. It will be able to house a large number of large experimental animals—and dispose of every ounce of their corpses without straining the local water treatment facilities. The NBAF will also house a highly classified intelligence operation tasked with providing early warning of emerging biological threats.

The NBAF lurks over the shoulder of the Kansas State University Biosecurity Research Institute (BRI), a place that has already assembled a dream team of scientists to study the panoply of infectious diseases that afflict plants and animals the world around, from African swine fever and Usutu virus to avian flu, from yellow fever to wheat blast.

Both the BRI and the NBAF had a powerful advocate in the former president of Kansas State University, retired four-star Air Force general Richard Myers. On September 11, 2001, Myers was vice chairman of the Joint Chiefs of Staff. On October 1 of that year, he was named chairman. He advised President George Bush and helped Defense Secretary Donald Rumsfeld through the invasions of Afghanistan and Iraq and the larger "global war on terror."

Myers was in the Pentagon in 2002 when US troops went into a cave in Afghanistan looking for Osama bin Laden—and instead found a list of pathogens, including human, livestock, poultry, and plant diseases that al-Qaeda was hoping to weaponize. Since returning to his alma mater as president in 2016, Myers has been helping to build Kansas State into a "Silicon Valley for biodefense."

"Agricultural threats tend to be overlooked even though food and feeding the world is critically important," said Myers, who retired from KSU leadership in 2022. "The world must be prepared, but we aren't."[28]

When the National Bio and Agro-Defense Facility opens on the Kansas State University campus, it will be able to work on Ebola and the other most frightening pathogens. "These diseases have been occurring naturally since the beginning of time," said Myers. "But what if people want to introduce this intentionally? It is a lot easier than finding fissile material and building something that goes bang. And the bad guys are going to be a long way away when anything happens."

The NBAF will also have the intelligence capabilities to try to spot bioterrorism against plants (destroying the enemy's crops is an ancient tool of war, but gene-editing creates new opportunities for doing so without detection) as well as biocrime (deliberately poisoning the food supply). There are experts working on agricultural cybersecurity, a growing threat as the United States automates more and more agricultural tasks. And there are even experts working on biological delivery mechanisms, including an entomologist who worries that flying

---

[28] Richard B. Myers (former chairman of Joint Chiefs of Staff and former president of Kansas State University), interview with authors, January 2019.

insects could be weaponized to deliver pathogens across the farm belt, slowly but undetected.

Just as the failure to anticipate 9/11 was a failure of imagination—that terrorists would turn commercial airliners into cruise missiles—Myers believes the same narrow thinking limits national security planning today.

"If I hadn't had that experience, I don't think I would have come here and reacted the way I did," he says of his commitment to turn KSU into the frontline defense against bio-, ag-, and chemical attack. "It's something we learned after 9/11: If adversaries are willing to die for the cause, what is the limit?"

After all, if hundreds of terrorists have proved themselves willing to blow themselves up in Afghanistan or Iraq or Pakistan or Indonesia, why not infect a few of their most fervent fighters with Ebola and book them a ticket from Africa to New York or Los Angeles? Or find a herd of cattle in the developing world sickened with foot-and-mouth disease—one of the deadliest to livestock, and therefore to a nation's economy—and simply swab the animal's nose with a handkerchief, stick that in a pocket—the virus lives for up to seven days—and fly to the United States, and tour any stockyard facility to sow economic catastrophe?

This is not a speculative scenario. In 1997, a group of New Zealand farmers were angry at their government for voting down efforts to use poisons to control a growing wild rabbit population. They smuggled in rabbit calicivirus RHDV1—a hemorrhagic virus that kills by causing organ failure—from the Czech Republic, where it is sold under license for pest control. The farmers, in what the government called an act of agro-terrorism, clandestinely spread the virus across a large portion of the island nation.[29]

Officials warn that an imported animal disease could work its way down the food chain and through the supply system before it is spotted, and the damage would be catastrophic. And such an attack does not require a high degree of technical know-how to disrupt the American economy with disastrous results.

This inability to know where an attack is coming from is known as "the attribution problem." It's a problem in cyberattack, when attackers spoof the digital credentials or take over the computers of innocent people to launch online assaults. Eventually, digital forensics can usually identify the culprit—but in the case of tracing the Russia hacking during the 2016 US presidential campaign, the process took months. Meanwhile, mistrust of government can fester, as does the opinion that the government is incompetent in dealing with emerging threats.

"The goal of terrorism is to create fear," Myers said. "Many people didn't fly after 9/11 out of fear. Plus, this kind of attack would erode trust in government. If you can't protect me, our food supply, our economy from this, what good are you?"

There are challenges in getting attention, and full funding, for the effort. In 2014, the CIA closed its office for collecting intelligence on biological threats. The Department of Agriculture's Plum Island facility for dealing with bio-threats

---

[29] Peter O'Hara, "The Illegal Introduction of Rabbit Haemorrhagic Disease Virus in New Zealand," *Revue scientifique et technique* (International Office of Epizootics) 25, no. 1 (2006): 119–123, http://boutique.oie.int/extrait/09ohara119123.pdf.

is crumbling and is not even certified to deal with the most lethal level of pathogens. There is a race underway to open the NBAF.

This illustrates a different kind of challenge to the warning and action machines: The correct threat assessment was made, and it was clearly heard, and the decision was made to set aside $1.2 billion to deal with it. But the tyranny of construction and testing will no doubt mean that the nation will be vulnerable, exposed to agra- and bio-threats, perhaps for a number of years, before the NBAF is fully online. To be sure, nobody wants to rush bringing a system online whose core mission is handling stuff that could end life or the food supply as we know it. But for this Manhattan Project, there is no broad sense of wartime urgency as there was for the first one.

According to officials at the NBAF office of communications, construction was completed in the summer of 2022, but the testing of all the facility's systems was expected to continue for months.[30] The best estimate offered by officials was that commissioning—meaning the facility is deemed safe for initial operations—would occur no earlier than the end of 2022, but that "full operation" still would be "some time away."[31] As systems are tested in this one-of-a-kind facility, efforts naturally are being made to ensure the highest standards are met. But even after the facility's commissioning, the NBAF's work will come online in stages as it undergoes testing

[30] Email exchange with authors and officials at US Department of Agriculture, conducted on background (no name) attribution, August 22, 2022.

[31] Michael Neary, "Officials Push Back Completion of NBAF Testing Process," *The Mercury*, August 24, 2022, https://themercury.com/news/local/officials-push-back-completion-of-nbaf-testing-process/article_4e099d5e-eae6-5caf-9c95-6e804e376f4b.html.

and validation, during what is called an operational endurance period. There also will be a "science preparatory stage," when experts will confirm the laboratory setup and equipment safety. And then the less-dangerous pathogens will be worked on in advance of those with a higher level of danger.

The officials stressed that little should be read into the transfer of NBAF from the Department of Homeland Security to the Department of Agriculture under a Memorandum of Agreement signed in 2019.[32] Although DHS and USDA certainly have different priorities, they share a common purpose, officials said, adding that the USDA has the broader and deeper technical expertise to be in charge of the NBAF.

"Even after commissioning is complete and the U.S. Department of Agriculture (USDA) takes ownership of the facility from the Department of Homeland Security (DHS), it will still take at least a couple of years to transfer the full science mission from the Plum Island Animal Disease Center to NBAF," the NBAF communications office said in a statement.[33]

"So there is a gap," Myers cautioned. "There is going to be a loss of momentum. It's like telling any potential enemy, 'Time out! Time out!' while the US readies for bio-agra threats. If you don't believe there is a threat, if you're not watching what's going on out in the world and paying attention, then you think, Why not sleep soundly tonight? I think we have to be ready for more of this."

---

[32] Memorandum of Agreement Between the U.S. Department of Agriculture Marketing and Regulatory Programs, the U.S. Department of Agriculture Research, Education, and Economics, and the Department of Homeland Security Science and Technology Directorate, June 20, 2019, https://www.usda.gov/sites/default/files/documents/usda-dhs-moa.pdf.

[33] Memorandum of Agreement Between USDA and DHS, June 20, 2019, https://www.usda.gov/sites/default/files/documents/usda-dhs-moa.pdf.

The Biden administration put its stamp on a strategy to counter biological threats and pandemics, releasing a new government strategy in October 2022 to update existing plans.[34] "The United States must be prepared for outbreaks from any source—whether naturally occurring, accidental, or deliberate in origin," the strategy states, and it requested Congress to set aside $88 billion over five years for biodefense and pandemic preparedness.

Another bold-faced person who has dedicated his late-career years to the problem is Eric Schmidt, best known as the longtime Google CEO and executive chairman. He now is committed to waking up the nation to unheeded threats in cyber and from artificial intelligence. Among them is how AI could be a catastrophically powerful force multiplier in biowarfare. He walked us through what he called a thought experiment to provoke, frighten, and perhaps inspire work to counter biological threats that he predicted will multiply at network speeds as advanced artificial intelligence becomes more widespread in society. The pathogens he imagines would be far more brutal than rabbit poison and easier to spread than by hiding them on a handkerchief.

"Say we build a large database of cellular information," Schmidt said to us from the headquarters of his Special Competitive Studies Project in the Washington suburbs.[35] "And we

[34] "Biden-Harris Administration Releases Strategy to Strengthen Health Security and Prepare for Biothreats," White House, October 18, 2022, https://www.whitehouse.gov/briefing-room/statements-releases/2022/10/18/fact-sheet-biden-harris-administration-releases-strategy-to-strengthen-health-security-and-prepare-for-biothreats/.

[35] Eric Schmidt (lead in Special Competitive Studies Project), interview with authors, April 2022.

build a set of tools that allow you to modify proteins and the way cells talk to each other. And it's generally available. So how many pathogens that would kill all of the human race can be generated by that?

"More than one," he said. "Maybe ten. Maybe a hundred. Maybe a thousand. Maybe one hundred thousand. Maybe a million. We can't even quantify the risk of a human-scale extinction," Schmidt warned. "OK. I'm saying that in order to be provocative. Maybe this won't kill you." (A beat for deadpan effect.) "Maybe it will just make you paralyzed."

There even is a name for this new kind of weapon and targeting: personalized warfare.

As these AI tools commercialize and become used broadly across society, "all scenarios that are bad will be tried," he said. "To me, the most obvious emerging threat is going to be this biological stuff," Schmidt concluded. "Because it's going to get real, and it's AI that will enable it. The biological one is really scary. We're not prepared for it, and we have some experience that we're not prepared for it."

The other significant problem with germs is that you don't need a bioterrorist to create a global health crisis. Looking back, Nicki Lurie couldn't help but reflect with sadness and apparent frustration.

"I would posit that one of the crazy things that happened early on in this pandemic is that people like my successor just never imagined that mother nature could be a worse actor than a bioterrorist," she said. "So, nobody heeded the warnings. At least in the US government, I mean, a couple people at NIH got their science right, you know they got the mRNA science going, but not as part of a national response."

CHAPTER 6

DIGITS

Defending at the Speed of Light

The idea that information could be networked and shared among large numbers of people goes back at least a century. Nikola Tesla conceived of a worldwide wireless system in the early 1900s. He never got it started. Others had different ideas about forming networks or connections among information systems. The problem was how to do it.

During the height of the Cold War, scientists and engineers puzzled over the idea of how to ensure secure communications in the event of a nuclear attack. The communications systems at the time had centralized switching networks. Think of an old-time operator sitting at a switchboard connecting one party to another with wires and plugs. The operators of old were in time replaced with automated switching, but the communications still ran through centralized hubs. Scientists and engineers knew you did not have to destroy the whole system to disrupt it. You simply had to disrupt the centralized hubs. Take down the hub, and you have taken down the system, even though most of the system is still intact.

This led them to think about a decentralized system, one in which communications could be broken into pieces or packets and flow from one source to another without necessarily having to travel through the centralized switching networks. In fact, the information could flow even if the key switches were destroyed. The information would flow in packets and be reconnected at the receiving end. You might think of this as being similar to how motorists leave congested highways to use secondary and tertiary routes only to reconnect to the main route after they pass the congested area. If there is a wreck at the interchange, motorists simply find ways to navigate around it. That is in contrast to a long train that must pass along fixed tracks and through central rail yards. Disrupt the movement through the central rail yards, and the system slows to a halt. Build secondary and tertiary routes, and the traffic can flow around it.

Paul Baran was a central figure, perhaps the central figure, in sorting out how to make a decentralized system work. Baran was a Polish-born engineer working for RAND in Santa Monica in the late 1950s and early 1960s. His family came to the United States in the late 1920s. He earned a bachelor's degree from UCLA by taking night classes. Baran was seized by a problem that was one of the hallmarks of the nuclear era:

> When I joined RAND in 1959, a glaring weak spot in our strategic forces command and control communications was a dependence on shortwave radio and the national telephone system, AT&T, both highly vulnerable to attack. H-bomb testing in the Pacific revealed that long distance short-wave (high-frequency) skywave transmission would be disrupted for several hours by a high-altitude nuclear blast. Computer simulations showed that weapons targeted at U.S. retaliatory forces

> would render long distance telephone communications service inoperative by collateral damage alone. While most of the telephone facilities would survive, the paucity of switching centers formed a dangerous Achilles' heel.[1]

This led to a series of innovative ideas that would be funded by the Defense Department. Ultimately, it was British physicist Donald Davies who independently created a concept he called packet switching that led to experiments in the late 1960s.[2] The result was the creation of the now famous ARPANET (Advanced Research Projects Agency Network), which was established to allow scientists and researchers to share information remotely. It got off to a rocky start, but within two years it blossomed into a high-speed electronic post office that allowed researchers to send packets of information across the globe. Thus began the modern World Wide Web.

It is interesting to note that Baran's ideas did not stop with decentralizing the communications network. By the mid-1960s, he was already thinking about a virtual department store where consumers could order products from their television screen.[3] He predated Amazon by nearly three decades.

---

[1] Willis H. Ware, *RAND and the Information Evolution: A History in Essays and Vignettes* (Santa Monica, CA: RAND Corporation, 2008), https://www.rand.org/pubs/corporate_pubs/CP537.html.

[2] Virginia Campbell, *How RAND Invented the Postwar World: Satellites, Systems Analysis, Computing, the Internet—Almost All the Defining Features of the Information Age Were Shaped in Part at the RAND Corporation* (Santa Monica, CA: RAND Corporation, 2004), https://www.rand.org/pubs/reprints/RP1396.html.

[3] "Paul Baran and the Origins of the Internet," RAND Corporation, https://www.rand.org/about/history/baran.html.

General Mike Hayden, who served as CIA and NSA director, compared the advent of the internet to other sweeping changes in global commerce and security. He refers to the centuries of European discovery. "That era, for all its accomplishments, jammed together the good and the bad and the weak and the strong in ways that had never been experienced before. What the Europeans got out of it was land, wealth, tobacco, and syphilis. Much of the rest of the world got exploitation of entire populations, global piracy, and global slave trade. We are at somewhat of an analogous condition now except that today's connectivity isn't at 10 knots with a favoring wind. It's at 186,000 miles per second."[4]

Hayden made this point in reference to criticisms that the World Wide Web was not being treated as a modern global commons for all to use and for none to dominate or control, and certainly not to weaponize as a theater for war. "It didn't take Stuxnet to make the cyber domain a very dangerous place."[5]

The cat-and-mouse nature of the modern internet goes back to the earliest days. In an enlightening and somewhat amusing story that later became a legendary book in cyber circles, Dr. Clifford Stohl, an astronomer and early computer security expert at Lawrence Berkeley Laboratory, discovered an anomaly on the laboratory's mainframe computer—a seventy-five-cent charge without a recorded user. This started Stohl on a chase that would lead to German intruders funded by the Soviet

[4] Michael V. Hayden, *Playing to the Edge* (New York: Penguin Press, 2016), 132.

[5] Hayden, *Playing to the Edge*, 132.

KGB. Among the computer systems that were compromised, a Defense Department database called Optimus, a NASA computer at the Jet Propulsion Laboratory in Pasadena, California, and computers used for nuclear weapons and energy research at Los Alamos, New Mexico, and the Argonne National Laboratory in Argonne, Illinois.[6]

This all took place in 1988 before most of us knew there was an internet. Clifford Stohl would go on to title his story *The Cuckoo's Egg*. It is widely read by students of the early cyber age.

We all know how Davies's packet switching changed our daily lives. It is worth considering its effects on how the national security machine conducts its business, collects information and intelligence, seeks to influence others—from individuals to mass publics—protects its own infrastructure, and conducts wars.

Any visitor to Washington, DC, New York, or any major city, for that matter, in the 1980s or 1990s would have seen hordes of couriers traversing the city streets at breakneck speed and conducting hair-raising maneuvers that would have impressed the savviest of Hollywood stunt persons. This, of course, was the age of paper and hard copies. Paper is what made the economy and government work, and it needed to move from building to building, block to block, and agency to agency. Couriers were the people who moved the paper across town and across agencies, but for the most sensitive messages,

[6] John Markoff, "West Germans Raid Spy Ring That Violated U.S. Computers," *New York Times*, March 3, 1989, https://www.nytimes.com/1989/03/03/world/west-germans-raid-spy-ring-that-violated-us-computers.html.

which were carried by government employees with padlocked briefcases. The Pentagon at the time was interlaced with vacuum tubes and messages were loaded into cylinders and bounced through walls and across ceilings. Staff officers spent their days "coordinating" decision papers by walking from office to office to collect initials and signatures from relevant officials. It was not uncommon for a staff officer to have his or her shoes resoled several times a year. The Pentagon's shoe repair store ran a thriving business.

By the mid-1990s, most federal government offices were wired with fiber-optic cable, and the coordination process began to take place virtually. Human contact became more virtual. Shoe leather lasted longer. Not everyone was thrilled. Some of the intermediate offices that were established to monitor and direct the paper flow found that higher-echelon offices skipped some of the middle organizations to find the actual experts and get speedier action. Such is life in a bureaucracy.

The idea of reaching deep into an organization for advice and expertise might have had its most exaggerated expression when Donald Rumsfeld showered the Pentagon with snowflakes—short memos, typically questions, that were directed to specific individuals in the Pentagon bureaucracy. Ironically, Rumsfeld was himself such an old-school person that his snowflakes arrived on eight and a half by eleven white paper. He used a Dictaphone to record his questions or observations and had an assistant type them, record a due date for a response, and send them through the Pentagon mail system. What many in the Pentagon didn't realize was that Rumsfeld was content to receive a handwritten response on the snowflake itself. His questions were entered into the electronic system and sent from desk to desk for detailed answers. The vacuum tubes were gone by the time Rumsfeld arrived, but paper still fell from above like snowflakes.

The advent of government online also made for leaks. The most notorious cases involved Army private Chelsea Manning (at the time known as Bradley) and Edward Snowden. Manning pled guilty to uploading a vast array of material to WikiLeaks in 2010. She was convicted and sentenced to thirty-five years in prison for her acts. Barack Obama commuted Manning's sentence four days before Obama left office.

Three years after Manning, in 2013 Edward Snowden would stun official Washington with leaks that detailed many ongoing intelligence operations. Snowden was twenty-nine years old, lacked a high school degree, and was employed by government contractor Booz Allen Hamilton. Snowden's plight would become bizarre as he sought refuge first in Hong Kong, then in a Moscow airport for forty days, in search of asylum status abroad. He ultimately remained in Russia, where he lives today.[7]

Of course, with new ways of sharing information, there came new ways of monitoring and stealing information. If it was once the job of intelligence to intercept mail and listen to telephone conversations, intelligence now had the job of monitoring the billions and trillions of messages traversing the internet. Much of this information sits in the open, there for anyone to see if only they know what they are looking for. Some of it is

[7] "Ex-Worker at C.I.A. Says He Leaked Data on Surveillance," *New York Times*, June 10, 2013, https://www.nytimes.com/2013/06/10/us/former-cia-worker-says-he-leaked-surveillance-data.html.

encrypted to keep it out of view and accessible only to those with specialized tools or modern-day decoders. And some lurks in hidden corners of the dark web where criminal and other illicit activity flourishes. Talk about finding needles in haystacks. Think of every laptop, tablet, and handheld as a potential haystack. All of them contain a needle of some sort. Few of them will have needles relevant to urgent national security decisions.

What all this led to was the creation of a whole new discipline within the intelligence community allowing it to gain access to treasure troves of information, scrape the data for relevant nuggets and clues, and piece them together with other sources of information to connect the dots or fill in the holes left empty by other more traditional sources. The new discipline is known as OSINT, or open-source intelligence.

It is not just the government that is in the business. OSINT is now big business in the commercial sector, too. Some businesses have built their tools for the government to use. Others have been created to serve a larger commercial or public need. With the ability to access public records, scan the internet for up-to-date information, and access commercial satellite imagery for sources, businesses have been entering this emerging field in large numbers. Anyone who watched Russian tanks burning along roadsides as Ukrainian troops sought to defend their cities, towns, and villages were undoubtedly being treated to various forms of OSINT. In the world of policymakers, it is not always the best information, but it is often the most rapidly available. So, it is the information that policymakers use.

Andy Roberts was the person trying to assemble a new organization for open-source intelligence and build new disciplines within the Defense Intelligence Agency. His job was not to remake the agency but to create a new node within it. He had to do so at a time when budgets were largely flat, which meant that anything new he needed had to come at the expense of something someone else wanted. Never a good place to be, certainly not within a large bureaucracy. He had to get access to data, including social media data, and people who knew what to do with the data. "So it was a combination of the tools, people, and the organization. Tools [used for scraping the internet] are great but only if you get the people to use them. And what we rapidly found out is that some tools were better than others. Not all of them had the data sets that we needed. That was the first thing we'd always check to see what they're offering."

Many of the tools and data sets were not owned by the government. More than a few sat in the private sector. Not everyone was interested in working with the US government, and certainly not everyone was interested in working with the intelligence community. "Some of them were not even keen on working with the government," said Roberts when we spoke with him.[8] Topsy, a tool used to scrape Twitter data and analyze information over time, "was something that we used back in 2014 that had a Twitter feed to it. And Apple bought them out and decided 'no way, no how are we going to deal with the IC.'" Apple paid $200 million for the company only to later shut it down.

This same problem would show up in a variety of areas. The celebrated cases of the US government and law enforcement wanting access to iPhone operating code is a good example. After the 2016 shooting in San Bernadino, California, in which a husband and wife couple shot and killed fourteen

[8] Andy Roberts, former Defense Intelligence Agency official, interview with authors, March 2022.

people, the FBI wanted access to the dead gunman's phone. A judge ordered Apple to comply, but Apple CEO Tim Cook said it would not. The situation was resolved, for the time, when the FBI found a private company to bypass the phone's security.[9] No matter the many good reasons offered about why access should be given, Apple always was reluctant to open the door, seeing it not as a sliver or a crack but a floodgate.[10]

In creating open-source intelligence, a new tool for the warning machine, the intelligence community also had to create a new type of analyst. Roberts described looking for "a collections discipline and getting folks trained in library sciences, data analytics, just a digital native by any means. They're the kind of people we started looking for, and how we brought them on board and put them through a pipeline of training to make sure that they got good tradecraft. And they can use the tools to go forth and do good things. And then organizationally, kind of give it a home, not just within DIA, but with the rest of the defense intelligence enterprise, then dovetail that with what others like CIA and DHS were doing out there."

This also came with something of a clash of cultures. Not just the new talent coming into the intelligence community, but also the older talent that was not ready to make room for the habits and likes of a new generation of talent. Think of it as white shirts and ties joining with polo shirts and flip-flops. "Some of the seniors, we had to convince them that, no, this

---

[9] Jack Nicas and Katie Benner, "F.B.I. Asks Apple to Help Unlock Two iPhones," *New York Times*, January 7, 2020, https://www.nytimes.com/2020/01/07/technology/apple-fbi-iphone-encryption.html.

[10] Vindu Goel and Nick Wingfield, "WikiLeaks Reignites Tensions Between Silicon Valley and Spy Agencies," *New York Times*, March 7, 2017, https://www.nytimes.com/2017/03/07/technology/wikileaks-silicon-valley-spy-agencies.html.

is the wave of the future, we're going to have to come to some proverbial middle ground to make sure we take an accounting of this . . . these people have tapped into something that we need to make sure we account for going forward."

Layered on this was the use of machine learning and artificial intelligence. "Machine learning and automation are very important," Roberts said, "especially in the last five years, I'd say. Just then, screening through the information you saw, like Twitter data, and it's literally like looking through multiple haystacks to find what you want to look for. So that's where automation became very important." Multiple haystacks. Maybe more like fields and fields of haystacks. Automation is providing the tools to sort the fields and haystacks in search of a few needles. That is, assuming the needles are lurking in the infosphere.

Roberts talked to us not just about timely information but also about the precision of the information. He discussed being able to geolocate where information was coming from and then to be able to verify the source. "Part of the tradecraft that some of the folks were learning is that reliance on the geolocational data is very important. It's telling everything from pattern of life to pattern of forces and so forth." "Pattern of life" is a military term to describe how intelligence, surveillance, and reconnaissance is used to understand the daily pattern of lives of adversaries—potential targets. "Pattern of forces" means how those units are deployed and positioned.

We saw the same ideas surfacing in our other discussions. Jason Matheny, the former IARPA director, talked of using geolocated data as an important source of warning. It was an IARPA team that broke the news about the 2014 Ebola outbreak in Guinea, West Africa, to US public health officials.

They did it by watching the automated detection of African news reports about an unidentified hemorrhagic fever. IARPA's big-data analytics have also been able to predict urban mass protests across the Middle East and North Africa far ahead of local station chiefs, diplomats, and local journalists by tracking purchases of vinegar, which blunts the effects of tear gas when poured on a bandana and held to the face, and of internet map searches to and from demonstration sites. And IARPA analysts uncovered early warning indicators of impending cyberattack by monitoring the black-market prices of malware.

The American military is well aware of the value of open-source intelligence and, during the war with Iraq, showed an unexpected creativity in harvesting and curating local media—even collecting rumors heard on the street, in the marketplaces, and on public transportation. During its deployment to central Iraq in 2003 and 2004, the First Armored Division assigned six US intelligence analysts, a pair of Arab American translators, and about a dozen local Iraqis to collect a range of OSINT and publish *The Baghdad Mosquito*—to capture, as it were, the "buzz" around town among the Iraqi population.

The publication, circulated among military commanders and senior American civilians at the occupation headquarters, was credited with having improved their information campaigns.

The reports of local press, gossip, and rumors collected by the publication "allow us to better understand what really are the concerns of the citizens on the streets of Baghdad," said then Brig. Gen. Mark P. Hertling, who was assistant commander of the First Armored Division during a deployment responsible for the security of Baghdad and central Iraq. "The feedback we received from *The Mosquito* was especially helpful in our design of a campaign countering the belief that all Iraqi police officers are corrupt and work contrary to the service of the citizens."[11]

Information tools can also be weapons of sabotage and war. Although some might have hoped for cyberspace to be a global commons—a place for communications and commerce—it has from the beginning been a place for competition and advantage. General Mike Hayden makes the point poignantly in his memoir: "The cyber domain has never been a digital Eden. It was always Mogadishu."[12]

We might think of it as the chaotic bar scene in the movie *Star Wars* or the Wild West.

The use of the Stuxnet worm to infect the software controlling Iranian centrifuges is a good example. Under the title "Olympic Games," the powerful cyber tool was developed under the Bush administration. George Bush wanted an option other than bombing Iran or giving Iran a free pass to nuclear weapons. The worm was initially developed under the guidance of US Strategic Command, which has responsibility for US nuclear weapons, but which also at the time was responsible for cyber planning. (US Cyber Command was established after the Stuxnet attack.) The program was inherited by the Obama administration, and after extensive review in 2010 the greenlight was given to activate the worm, which caused hundreds of Iranian centrifuges to spin out of control. Knowledge of the program might never have leaked out had the Stuxnet worm

---

[11] Thom Shanker, "The Struggle for Iraq: Civil Society; U.S. Team in Baghdad Fights a Persistent Enemy: Rumors," *New York Times*, March 23, 2004, https://www.nytimes.com/2004/03/23/world/struggle-for-iraq-civil-society-us-team-baghdad-fights-persistent-enemy-rumors.html.

[12] Hayden, *Playing to the Edge*, 132.

not made its way out of Iran's nuclear facilities to infect computer systems across the globe.[13]

Stuxnet was the first known use of a cyber tool to produce a kinetic effect—that is, the worm infected the software "brain" that caused the centrifuges to spin out of control. There was never an expectation that Stuxnet would sideline Iran's nuclear program. The hope was that it would slow it down, buy time, and perhaps give sanctions and negotiations more time to work. Before the attack was launched, there was an extensive effort to determine what other type of damage might result. The target was the centrifuges, not the local area or hospitals or grocery stores, or the electric power generation plant, for that matter. The desire was a localized effect to buy time. Even still, the worm escaped, which was an embarrassment and a warning. In a world where things are moving at 186,000 miles per second, things can get out of control.

The thing about systems is that few of them stand alone in a connected world. Systems are connected to systems, which are connected to still other systems. It is a gross exaggeration to say everything is connected. Some systems are reasonably well protected. The technology industry knows where its jewels are hidden, and it has taken extraordinary measures to protect them. So, too, with the banking and financial industries, which have learned to become technology industries of their own. But not all systems are protected, including some of the government's own systems. One only needs to recall the 2015 Office of Personnel and Management data breach. Sensitive information for over twenty million federal workers, federal

---

[13] For an excellent, contemporaneous summation, see David Sanger, *Confront and Conceal: Obama's Secret Wars and Surprising Use of American Power* (New York: Crown, 2012).

contractors, and family members—yes, twenty million—was swept off US government servers onto servers presumably in Beijing. Names, addresses, prior residences, places of birth, family members, marriage information, social security numbers, civil and criminal records, and a whole lot more.

Of course, every consumer who has received notice that credit data has been compromised is well aware how vulnerable unprotected data can be.

If this was 2012, we would have said the most significant cyber threat was that cited by Leon Panetta, the defense secretary, in his speech aboard the aircraft carrier *Intrepid* docked in New York. Panetta warned that the United States was facing a "cyber Pearl Harbor"—that foreign hackers could dismantle the nation's power grid, transportation system, or financial networks.[14]

If this was 2014, we might have said that the most significant cyber threat to the United States was something like North Korea hacking into corporate computer systems, such as Sony.

And if this was 2016, then obviously we would say that the gravest threat to the nation's cybersecurity was Russian hacking to disrupt American elections and undermine our democracy.

And if this was 2021, the biggest threat would be cybercriminals, likely abetted by a nation-state, Russia, using ransomware to shut down a critical pipeline and send gas prices spiraling upward and the economy spiraling in the opposite direction.

---

[14] Elisabeth Bumiller and Thom Shanker, "Panetta Warns of Dire Threat of Cyberattack on U.S.," *New York Times*, October 11, 2012, https://www.nytimes.com/2012/10/12/world/panetta-warns-of-dire-threat-of-cyberattack.html.

Today, which is it? Or something else? Mieke Eoyang, the Pentagon's top official for cyber policy, challenges many of the terms, and even assumptions, underlying the current debate on digital threats.

"Talking about some of these things as weapons and arsenal and analogies that we use from traditional warfare may or may not be accurate when it comes to cyber," said Eoyang, who was named deputy assistant secretary of defense for cyber policy after a long career as senior Capitol Hill staff member. "And in many cases, they're actually not accurate and they're not helpful for people understanding the debate."[15]

Although much of the debate over cyber war, cyber threats, and cyber policy has drawn language from nuclear arms and nuclear arms control, Eoyang said the analogy is far from accurate.

"A nuclear weapon will make a space unlivable and deny access to it for generations," she noted. "Cyber weapons may be decisive in the sense that they change things at a particular time that make a difference to an adversary's calculus." Cyber weapons may make a centrifuge spin out of control, but they don't flatten cities. It is not clear how temporary some of the effects might be—cutting off the power to a nuclear power plant could have lasting effects—but they won't produce the piles of rubble that Russian bombs and artillery shells did in Grozny or Mariupol.

---

[15] "Mieke Eoyang, Deputy Assistant Secretary of Defense for Cyber Policy" (transcript), Cyber Media Forum, Project for Media and National Security, George Washington School of Media and Public Affairs, October 20, 2021, https://nationalsecuritymedia.gwu.edu/project/mieke-eoyang-deputy-assistant-secretary-of-defense-for-cyber-policy/.

She cited the Colonial Pipeline, taken down in 2021 by a ransomware attack mounted by a criminal group that may or may not have had the blessing, however tacit, of Russia. "As we saw with Colonial Pipeline, eventually people who are determined to reconstitute and continue their operations will do so," Eoyang said. Gas stations closed for a time along the East Coast of the United States. Anywhere that had gasoline also had long lines of anxious, sometimes angry drivers waiting to fuel up. "So it is a time-limited effect. People come back online. So, we have to think about the impact of these things not as the kind of kinetic weapons that leave things a smoking crater for a long period of time, but about the will to reconstitute."

A focus on cyber warfare, she noted, also clouds focused thinking on the other ways an adversary or rival might use digits in a hostile way, including espionage, theft of intellectual property, and information warfare.

Cyber responsibilities were once the purview of US Space Command in Colorado Springs. When Space Command was disestablished in 2002, cyber responsibilities were transferred to US Strategic Command in Omaha, Nebraska. By 2004, the Joint Chiefs of Staff had declared cyberspace a domain of conflict, and in 2010 then Secretary of Defense Robert Gates called for the creation of a cyber command, which would be collocated with the National Security Agency in Fort Meade, Maryland, just outside Washington, DC, but it would still be subordinate to Strategic Command. In 2018, President Donald Trump designated US Cyber Command as the eleventh combatant command.

Even still, not everyone is thrilled with how cyber policy and cyber tools are being developed and employed. Sue Gordon and Eric Rosenbach are veterans of the national security machine. Gordon was deputy director for national intelligence. Rosenbach was chief of staff to Secretary of Defense Ashton Carter. In a 2022 article, they recall how a senior Russian officer belittled the US organizational approach: "One uses information to destroy nations, not networks. That's why we're happy you Americans are so stupid as to build an entire Cyber Command that doesn't have a mission of information warfare!"[16] This exchange took place in 2013, perhaps a harbinger of what the Russians were planning in the 2016 American elections. They recount that, time after time, lack of proper organization and planning led to a lack of appropriate options after the Russian takedown of the Ukraine power grid, the nonstop Chinese theft of US intellectual property—valued between $200 billion and $600 billion per year—and the hack of the Office of Personnel and Management database, to name just a few. They applaud the designating US Cyber Command as a stand-alone command, but note that Trump himself undermined the development of coherent policy by his "bizarre genuflection toward Putin."

Gordon and Rosenbach argue that Obama's passivity and Trump's inconsistency left President Joe Biden with a mess. The authors' recommendations include fighting fire with fire—using cyber to fight cyber, as in hacking back—but also to draw on other strengths, including other economic and military strengths to deter cyberattacks. They also argue that the best deterrent is not one of punishment after the fact, but denying access before the fact. That means protecting important data and facilities and instilling better discipline on the part of the

---

[16] Sue Gordon and Eric Rosenbach, "America's Cyber Reckoning," *Foreign Affairs*, January/February 2022, 13.

people operating the systems. They want to provide citizens with legal recourse against companies that fail to protect their data, though they aren't clear whether that includes the government itself. They also want to make the Cyber and Infrastructure Security Agency in the Department of Homeland Security the centerpiece of domestic cyber policy by giving it broader authorities. Like other areas we have seen, responsibilities are still too fractionated. In the world of government speak, they want to find a "belly button"—a person who is in charge—if something goes wrong. They also want to see Cyber Command become more like the nimble Special Forces Command, focused more on speed of development and speed of action rather than lumbering organizational approaches. They want martial arts, not blocking and tackling.

Gordon and Rosenbach seem to agree with Mike Hayden that the cyber domain is Mogadishu, not Eden. But they argue that cyber policy doesn't have to be the chaos of Mogadishu. More to the point, they are among the people who want to fix the larger national security machine, who believe that for all the tinkering we still have not created the organizational tools to contend with the problems embedded in the larger cyber domain. They are not alone in thinking we are still driving an old Chevy when we need a modern Tesla.

In December 2021, Congress created the position of National Cyber Director. It is a position the Biden administration didn't want but that the administration filled out of recognition that the subject is far too important and the existing machinery is not up to the task of dealing with the cyber challenges that have emerged. The job of the coordinator is to help bring coherence to the dozens of agencies that have responsibilities for cyber security and spend upward of $100 billion a year on information technology and cyber tools. The coordinator's job is to bring a modicum of order to what is otherwise a fairly chaotic field—one that might be best described as having too many cooks in the kitchen. Among the major tasks will be to bring more coherence to the ways the federal government ensures the safety and security of the larger public infrastructure—electricity, water, hospitals, and transportation. With so many cooks in the kitchen, the rules and guidelines can be confusing and even contradictory.[17]

That a new role has been created in the White House has both pluses and minuses. On the plus side, there is recognition that no one is in charge and the fractionation of responsibility has led to confusion and slowness in creating a coherent federal approach to the making of policy. That said, there is already a cyber coordinator on the National Security Council staff, so now there are two coordinators who need to make sense of their own jobs and the larger federal approach. On the minus side, whenever a new coordinator position is established in the White House, there is the risk that the urgent will displace the important. Coordinators are established to coordinate, not to direct or control. Authority still flows from the president to the cabinet officers, and coordinators are only as influential as they are able to capture the president's ear. There is never enough time in a day, or a week, or a month for a president to take up a coordinator's cause. This is not to suggest that all in the federal sector will be in active defiance of the coordinator. But each

[17] Eric Geller, "The First National Cyber Director Has Big Plans to Toughen U.S. Digital Defenses," *Politico*, August 30, 2021, https://www.politico.com/news/2021/08/30/chris-inglis-cyber-attacks-507021.

department and agency has its own priorities and is motivated more by the priorities of the department or agency than it is by a coordinator in the White House. To the extent the coordinator has control over spending, then priorities might change, but that is a difficult challenge to take on and is further complicated by relationships among departments and the committees that oversee them in Congress.

Cyber is now a very crowded field in which government and private-sector functions overlap across all areas of modern life. Think of a kitchen with many stoves and many cooks. Any number of pots could boil over at a given time. The departments and agencies need to manage those stoves and keep the pots from boiling over. They have to confer with one another and their partners in the private sector as they are doing so. The energy sector will want to deal with the Energy Department, the health sector with Health and Human Services, the transportation sector with the Transportation Department, and so on. A coordinator can help set priorities and bring greater coherence to the overall effort, but the coordinator is not in charge of the controls that keep the pots from boiling over. That means we should expect more chaos in this crowded field. It also means we almost certainly have not seen the last of the reorganization plans.

Outside the government there is a wide circle of people thinking about the future of the digital age. Jared Cohen is one of them. He has a knack for being a decade ahead of everyone else. Not just in what they are thinking but also in what they are doing. He graduated from Stanford in 2004, having already worked at the Pentagon and the State Department. At age twenty-four, he joined the Policy Planning Staff at the State Department—the office founded by George Kennan at the dawn of the Cold War—and worked for both Condoleezza Rice and Hillary Clinton. At twenty-nine, he was hired by Eric Schmidt at Google to form Google Ideas, which later became Jigsaw, an incubator focused on supporting internet freedom. He now works for Goldman Sachs.

Cohen lives at the intersection of policy and technology, and cyber policy is very much on his mind. He talks about cyber policy in interesting ways. Strategists are always in search of weaknesses. That's the nature of the strategist's business. Build on strength; find and exploit weakness. Cyber strategists are in search of actual human errors. Not just weaknesses or vulnerabilities, but mistakes. "What's interesting about cyberattacks is they oftentimes require human error in order for them to land right. Unlike a heat-seeking missile, where you can direct it towards something warm, with cyber you direct it towards carelessness."[18] It is like searching for a chink in the armor. Sometimes the designers know the vulnerability exists, and they leave the door open to see if others try to get in. Other times, no one is aware of the vulnerability until someone has walked through the door and cleaned out the house.

In his own way, Cohen goes on to describe why it is difficult to compare cyber weapons to other weapons of war, particularly nuclear weapons. He described to us the difficulty of using weapons whose source can be hard to ascribe and whose effects might simply be a dud. He talked of a cyberattack that would land with a thud if it came up against a well-protected target. What do you do with a weapon that could have a devastating effect but that lands with a thud. Nothing happens? And the attacked party now likely has the code, the unexploded

---

[18] Jared Cohen (Google Jigsaw), interview with authors, March 2022.

weapon, if you will. Do you fire back? Do you tell the world that you were subject to attack and that the enemy missed? "All of this is a long way of saying you have no deterrence in this space. And nobody's figured out how to have deterrence in this space. That, I think, is the biggest problem," Cohen explained to us. At least deterrence that comes in the form of threats of punishment. His world is one of denial because it is too hard to attribute an attack in ways a president would be willing to punish an offender, either with cyber or a more traditional military response. In the world of cyber, better to deny an attack than deal with the consequences of an attack.

Cohen seems to share Hayden's view of Mogadishu, but he wasn't always of that mind. He thought the United States and China might have had a chance to settle on a set of global norms. He did not go so far as to think these would be recorded in a treaty or formal set of accords, but he did think some form of global norms might have been possible. "I've actually always thought there might be room for the US and China to sort of engage on this; I feel less this way now. But in simpler times, I thought the 2015 Obama–Xi summit on this actually went better than I expected." It was here that the United States and China agreed to control cyber theft and piracy. "You know, I was skeptical that a meeting would reduce the volume of attacks, but it did for at least a year. We saw that at Google."

Cohen went on to say, "I thought that it could be interesting, piggybacking on that, I thought we had this window where the US and China and other countries coming together and agreeing on a list of targets that were basically off-limits—ventilation systems and tunnels, water supplies, things that will lead not just to loss of life, but real suffering." If there were to be an agreement on what is "off-limits," then there might be the beginning of a framework for how to think about deterrence. It starts with clarity on what is off-limits. It would not necessarily limit non-state actors, but it would begin to establish a set of international norms.

"I want to make sure that if we stand up and say, 'if you do this to us, we're going to do that to you,' " said Cohen. "They understand what we're talking about. And then if they do, if they say the same thing to us, we understand what they're talking about.... Anything we can do to remove ambiguity in what is already a perpetual state of cyber conflict. Anything we can do to take ambiguity off the table moves us closer to deterrence," Cohen observed.

It didn't happen, and it's not likely to happen now. Not with China, at least. Russia, either.

We took Cohen back to the world of humans and human mistakes. We asked him about his worries, the problems that had not been solved and that could become calamitous. He began with where he was less worried. Although the banking system is often one of people's Pearl Harbor scenarios, this wasn't a top concern for Cohen. He isn't Pollyanna-ish, but he believes the banking system has become enough of a technology hub of its own that it has reasonably effective standards, protocols, and discipline in place. That is not to suggest parts of the system could not experience disruption or losses, but rather the system as a whole is relatively robust.

We kept probing. What about this? What about that? Then we arrived at public utilities. The tone of his voice changed a bit, and his reaction was crystal clear. When we asked, "Where would you put public utilities on your list of worries?" Cohen responded, "Pretty high. I actually think the level of security that exists with public utilities is pretty appalling. I can't figure out why. I mean, it's not clear, I'm shocked that something hasn't happened yet."

A February 2021 hack at a Florida water plant is exactly the type of attack Cohen had in mind. A plant operator observed that his cursor on his computer was outside his control. The hacker then went on to add a hundred times the normal level of sodium hydroxide—better known as lye, one of the main ingredients in drain cleaner—to the water supply. Fortunately, the plant operator observed the hack and returned the sodium hydroxide to normal levels.[19] On this occasion the hacker was stopped in his tracks. Who knows next time.

Cohen went on to say, "I actually think the more likely scenario is not a state-sponsored attack but something that's more innocuous that just kind of spins out of control. Because it's not just that these public utilities don't have the right defenses. They have workforces that are much more prone to human error." Perhaps a hacker fiddling with additives to the water supply?

Back to human error. To look for weakness in the cyber realm is to look for human error.

Cohen went on: "Let's take an electric grid, I think it's going to be disproportionately targeted. It's built on antiquated systems. The systems are deliberately decentralized, which makes them harder to update and so forth. And the workforce is just, say, not among the most digitally literate workforces when it comes to this stuff. It's a bad combination."[20]

---

[19] Jenni Bergal, "Florida Hack Exposes Danger to Water Systems," Pew Charitable Trusts, March 10, 2021, https://www.pewtrusts.org/en/research-and-analysis/blogs/stateline/2021/03/10/florida-hack-exposes-danger-to-water-systems.

[20] Jared Cohen, interview with authors, March 2022.

There's a further wrinkle. Close cyber watchers have been anticipating the advent of quantum computing for years. It will offer computational advantages that could have only been imagined in the past. It could also provide users with the ability to break the encryption that has been the backbone of digital safety. Without reliable encryption there is no digital security.[21] It is a huge challenge, and the best minds are working on the problem, including the best minds in China. Whether the breakthroughs in quantum computing convey a first mover advantage—the first to gain the capability will have a huge advantage—is being hotly debated, but those pursuing the capability want to get there first. The competition over quantum is being hotly pursued.

If cyber remains a constant worry because of the lawless Wild West nature of the threat, there is a particular element of the cyber domain that causes special worry. We have been speaking of the world of digits. The part of the cyber domain that has people especially concerned is a world where digits have brains, or what is more commonly known as artificial intelligence.

Ever since the advent of modern computing, humans have been enamored with the idea of computers with brains, not just computers that can calculate but computers that can reason on their own. There were discussions of artificial intelligence in the 1950s, and with each decade these talks have taken on more and more importance. Only in the last decade or two has the promise of artificial intelligence become real. Most remember

---

[21] Michael J. D. Vermeer and Evan D. Peet, *Securing Communications in the Quantum Computing Age: Managing the Risks to Encryption* (Santa Monica, CA: RAND Corporation, 2020), https://www.rand.org/pubs/research_reports/RR3102.html.

the AI HAL from the movie *2001: A Space Odyssey*. The world is now populated with HALs.

"So, [right now] AI is 10 percent of what it's going to be," explained Eric Schmidt, former CEO of Google, who spoke with us from his offices in Northern Virginia. The one-time businessman is now focused on philanthropy and advising the government on how to prepare for the world of AI challenges.[22]

Schmidt has written extensively of the promise of artificial intelligence. In a book he authored with Henry Kissinger and Daniel Huttenlocher, they describe the promise of AI, the impact it will have on our lives, and how it will shape interactions among individuals and across nations.[23]

In our discussion with Schmidt, he focused on the more treacherous side of AI. He was not describing how AI will beat the world's chess masters or how it might be used to compose music. The world Schmidt was describing was not one in which every computer has the potential for creative thought, either. Rather, he was focused on how a handful of computers will possess the ability to produce massive good or massive harm. More than once, he said, "We're in the 10 percent. There's 90 percent ahead of us."

Schmidt described for us a future when computers will act like humans and will decide for themselves what they want to work on. "At some point, some people are saying twenty years, but not soon, there will be systems that you can talk to, that will talk back to you, and unlike current systems, they'll be able to choose their objective function. In other words, they'll decide what they want to work on."

---

[22] Eric Schmidt (former Google CEO), interview with authors, April 12, 2022.

[23] Kissinger, Henry, Schmidt, Eric, and Huttenlocher, Daniel, "The Age of AI: And Our Human Future," Little, Brown, and Company, 2021.

Schmidt quickly corrected himself. " 'They' is a misnomer. 'They' are computers, not humans. It will decide what it wants to work on."

He said there will not be a lot of these computers because they will be very expensive to build and operate. The systems he is describing won't be the result of creative minds in a garage somewhere building a new computer. This won't be the remake of Steve Jobs and Steve Wozniak. They will be the purview of states.

He went on to describe a chilling scenario for us: "So let's say there'll be ten computers. I'll make up a number. And now say there's two or three in China, two or three in the US, one in Israel, one in Europe, maybe one in Russia, who knows?"

Schmidt went on, "What will be the nature of access to that computer? Because with this kind of computer, you could say, 'I want to know, I want you to invent a way to kill one million people who are not my race.' Okay, so you say, 'Well, it's obvious that what you should do is you should have a front end that doesn't allow that query.' Okay, so we'll ban that query. Okay. But since it's a general conversational system, the average person can keep trying until they get it to actually understand what they want and then it will do it."

Keep asking until the computer responds, no matter how it was designed. Sounds like another matter of human error, not computer error.

"So you have a proliferation problem as follows. You have to figure out how to keep this capability out of the hands of bad people and you also have to deter its use by bad governance. So, let's assume, for example, that North Korea steals the China one, which it does through some relatively easy mechanism. They managed to get it into North Korea, they managed to turn it on, and now they're busy trying to use it. Well, what is the deterrence architecture to prevent them from turning it on? These questions are completely unexplored."

In a way, Schmidt is replaying the early years of the nuclear era, but he wants to do so with much more deliberate thought. It was nearly a decade into the nuclear era before the discipline of nuclear deterrence emerged, before we had terms like *second-strike stability* and *MAD*, before *Dr. Strangelove* became part of the lexicon.

Schmidt sees a potentially frightening future emerging, and he wants to do the hard work now before we stumble into the abyss. Humans are devious enough. Schmidt is not eager to turn computers loose on the darkest human thoughts.

"They're more powerful, they have more data, they can begin to do things that we haven't figured out how to prevent yet."

Schmidt is sending a warning. The machine is not ready for the machine.

Eric Schmidt isn't the only person sending a warning. In 2018, Kai-Fu Lee wrote a book titled *AI Superpowers*. Lee grew up in Taiwan and has deep roots in the technology industry and Silicon Valley. He headed Google China until Google decided to leave China. He wrote *AI Superpowers* to sound his own warning, and it is a dire one. He doesn't dispute Schmidt's views on the state of play of AI in the modern world—that we have seen only 10 percent of the potential—but he sees China bringing particular advantages to the competition.

Lee compares AI with the early days of electricity. Once Edison harnessed electricity, there was no telling what electricity would do to change modern life. Bringing light to darkness was one thing. Bringing power to machines was another. Little would Edison know in the early days of electricity that it would be used to refrigerate food, wash and dry clothing, move people from the ground to the top of the tallest buildings in the world, move people across cities in trains, regulate traffic, and ultimately provide the energy for human-like thought.

He goes on to discuss the larger state of play in the world he describes as "deep learning." He begins by noting this was once a distinct US advantage. The big ideas in the world of AI were coming out of the United States and Canada. That was during the time that he calls the period of discovery. But he goes on to say we are no longer in a period of discovery. We are now in a period of implementation. "During the age of discovery, progress was driven by a handful of elite thinkers, virtually all of whom were clustered in the United States and Canada. Their research insights and unique intellectual innovations led to a sudden and monumental ramping of what computers can do."[24]

But now we know what computers can do, and there has been a shift in the larger field. "Today, successful AI algorithms need three things: big data, computing power, and the work of strong—but not necessarily elite—AI algorithm engineers."[25] In each of these three areas, he believes China has a distinct advantage. In prior technological revolutions, advantage conveyed to the inventors, the first movers who possessed the skill and the know-how and put it to use. Lee sees no such advantage in this revolution. He notes with apparent amusement, "When asked how far China lags behind Silicon Valley in artificial intelligence research, some Chinese entrepreneurs

[24] Kai-Fu Lee, *AI Superpowers: China, Silicon Valley, and the New World Order* (New York: Houghton Mifflin, 2018), 13–14.

[25] Lee, *AI Superpowers*, 14.

jokingly answer 'sixteen hours,' the time difference between California and Beijing."[26]

With these pieces in place, Kai-Fu Lee goes on to make a chilling prediction: "China will soon match or overtake the United States in developing and deploying artificial intelligence." He goes on to say, "That lead in AI deployment will translate into productivity gains on a scale not seen since the Industrial Revolution."[27]

He then talks about the implications for the US workforce: "I predict that within fifteen years, artificial intelligence will technically be able to replace 40 to 50 percent of the jobs in the United States."[28] He sees similar impacts in China but concludes job losses will come first to the United States. He is not predicting 40–50 percent unemployment because interventions of various forms will slow many of the losses. But the losses will be significant nonetheless.

Not everyone buys Lee's analysis and projections. Some who disagree believe China's authoritarian system will be its undoing. It will become brittle when it needs to be resilient. It will not adapt well to the social changes that are inevitably coming to China's polity. China's demographics alone could prove to be its undoing. Demographers note that China's population will decrease by half—yes, half—over the next seventy-five years.[29] Others suggest it might even be faster, perhaps within forty-five

---

[26] Lee, *AI Superpowers*, 87.

[27] Lee, *AI Superpowers*, 18.

[28] Lee, *AI Superpowers*, 19.

[29] Stuart Anderson, "China's Population to Drop by Half, Immigration Helps U.S. Labor Force," *Forbes*, September 3, 2020, https://www.forbes.com/sites/stuartanderson/2020/09/03/chinas-population-to-drop-by-half-immigration-helps-us-labor-force/.

years.[30] But those steeped in the field of AI have read Kai-Fu Lee and pay attention to him. The stalwarts believe America's chances are better than he suggests, but they do take heed of his warning. They are doing their best to bring this to the attention of America's leaders and the public at large.

One last reflection. We have written elsewhere of Sputnik moments. Kai-Fu Lee writes of China's Sputnik moment. It was when teenager Ke Jie was pitted against Alpha Go, an AI powerhouse supported by Google. Ke Jie was the premier player in China of the ancient game of Go, which is played on a nineteen-by-nineteen lined board with stones. As Kai-Fu Lee describes it, Ke Jie was not simply bested by a thinking machine, he was thoroughly dismantled. This occurred in 2017. It created a realization among China's leaders that AI was a competition that China had to win. There has been no looking back since.[31]

Eric Schmidt, Bob Work, Jason Matheny, and other technologists served on the National Commission on Artificial Intelligence. They issued their report in late 2021, over six hundred pages including appendices. The report has the feel of many commission reports—lots of facts and figures, numerous recommendations. The report is full of warnings from serious people and a call to action—to America's leaders and the American people. The report focuses on talent, technology, intellectual property, and many other topics.

---

[30] Stephen Chen, "China's Population Could Halve Within the Next 45 Years, New Study Warns," *South China Morning Post*, September 30, 2021, https://www.scmp.com/news/china/science/article/3150699/chinas-population-could-halve-within-next-45-years-new-study.

[31] Chen, "China's Population Could Halve."

What the report is really about is waking America from its slumber.

> We are fortunate. The AI revolution is not a strategic surprise. We are experiencing its impact in our daily lives and can anticipate how research progress will translate into real-world applications before we have to confront the full national security ramifications. This commission can warn of national security challenges and articulate the benefits, rather than explain why previous warnings were ignored and opportunities were missed. We still have a window to make the changes to build a safer and better future. The pace of AI innovation is not flat; it is accelerating. If the United States does not act, it will likely lose its leadership position in AI to China in the next decade and become more vulnerable to a spectrum of AI-enabled threats from a host of state and non-state actors.[32]

The report lists a host of concerns and remedies, but one particular observation stands out: "The best human operator cannot defend against multiple machines making thousands of maneuvers per second potentially moving at hypersonic speeds and orchestrated by AI across domains. Humans cannot be everywhere at once, but software can."[33]

---

[32] Eric Schmidt et al., *Final Report: National Security Commission on Artificial Intelligence* (Washington, DC: National Security Commission on Artificial Intelligence, 2021), 19, https://www.nscai.gov/wp-content/uploads/2021/03/Full-Report-Digital-1.pdf.

[33] Schmidt et al., *Final Report*, 24.

They want the software to be in American hands, and they know the United States can lead. They also know the machinery isn't yet in place for this to happen and that time is short to get it in place.

Their concern?

> The U.S. government still operates at human speed, not machine speed. Adopting AI requires profound adjustments in national security business practices, organizational cultures, and mindsets from the tactical to the strategic levels—from the battlefield to the Pentagon. The government lags behind the commercial state of the art in most AI categories, including basic business automation. It suffers from technical deficits that range from digital workforce shortages to inadequate acquisition plans, insufficient network architecture, and weak data practices. Bureaucracy is thwarting better partnerships with the AI leaders in the private sector that could help. The government must become a better customer and a better partner. National security innovation, in the absence of an impetus like a major war or terrorist attack, will require strong leadership.[34]

The commission wants to build a warning and action machine that looks and drives more like a Tesla than the '47 Chevy that is parked in the garage. It is not about adopting technology or hiring digitally savvy personnel. It is about taking on a new mindset. It's like the earliest days of the nuclear era. The country's

---

[34] Schmidt et al., *Final Report.*

best minds are needed to support an urgent national cause. And the institutions that surround them need to adapt so they can contend with new problems at the speed at which they are coming. There is very little that is slow in the world of cyber or AI but the institutions we have created to deal with the threat. General Mike Hayden was right, the problems are coming at us at 186,000 miles per second. The old Chevy, no matter how sleek or shiny, simply wasn't built for that. Best to leave it in the garage.

There is still the question of a cyber Pearl Harbor. Leon Panetta warned of it a decade ago. In a later conversation, Panetta updated his warning on a cyberattack that was a cyber Pearl Harbor, noting that viruses can operate in a far stealthier manner than even the Japanese sneak attack that brought the United States into World War II. "I worry that the sophisticated viruses are already implanted, and that we haven't developed the capability to find those viruses and to defend ourselves, once they're activated," Panetta said.[35] "Every one of those viruses can operate as a separate missile coming at the United States, taking out our power grid, taking out our financial systems, or taking out government systems. We are always worried that terrorists could get a hold of a nuclear weapon. I think my greatest fear is that terrorists who don't give a damn will get a hold of a sophisticated virus and not be afraid to use it."

So, what to make of the issue?

A Defense Department cyber official suggested it may have already happened. We pressed him on the question, "Why

---

[35] Leon Panetta (former secretary of defense), interview with authors, June 2019.

haven't we seen it?" His answer was intriguing. "Maybe you have, and you just don't understand what it is you were seeing." He referred to an observation made over a decade ago by General Keith Alexander, who was then director of the National Security Agency. Alexander was commenting on the disruptive quality of cyber. He was comparing it with the potential for destructive uses of cyber, how digits could be used to destroy as well as disrupt.

In the course of this discussion, Alexander made a more profound observation. In referring to the ongoing theft of intellectual property, then estimated at $250 billion a year, he observed this is the "greatest transfer of wealth in human history."[36]

Had this been Thomas Jefferson's era, the president would have likely dispatched the Marines as Jefferson did to deal with the Barbary pirates off the coast of Tripoli. But this theft was different. It wasn't perpetrated by a lawless group operating from an ungoverned area. Much of it was coming from China, which is anything but lawless. In its quest to get ahead, it was willing to steal for advantage, and they found the door unlocked and the wallet sitting on the table. The cyber army the United States built was not organized to enforce the law. And, for a lengthy period of time, it was not clear the private sector in the United States was looking to a law enforcement arm for help. The US private sector still thought there was business to be done in China, and it was willing to suffer losses

---

[36] Josh Rogin, "NSA Chief: Cybercrime Constitutes the 'Greatest Transfer of Wealth in History,'" *Foreign Policy*, July 9, 2012, https://foreignpolicy.com/2012/07/09/nsa-chief-cybercrime-constitutes-the-greatest-transfer-of-wealth-in-history/.

for what it thought would be much larger gains. That changed when Xi Jinping took a decidedly nationalist and authoritarian turn as he ascended to control of the Chinese Communist Party. But, by then, the damage had already been done.

Wars of conquest are usually fought with armies. This one is being fought with digits. China is not the only culprit in stealing intellectual property, but it is a dominant one. It has acquired enormous wealth without ever firing a shot.

CHAPTER 7

DRONES

An American Monopoly Lost

It reads almost like a made-for-TV script. A dark and rainy night, three a.m. in late January 2015. An off-duty government employee, who later admitted to having had a couple of drinks, stood on an apartment balcony in downtown Washington, DC, and launched a two-foot-by-two-foot "quadcopter" that costs just a few hundred dollars. Investigators later determined that the drone's "pilot"—and we are using the title loosely—lost control, whether owing to wind or trees or signals blocked by buildings as the aircraft flew out of line of sight.[1] The guidance system was unable to penetrate or go around nearby obstructions.

So, he went to bed.

The gyrocopter went down on the White House grounds.

The Secret Service went on high alert.

The incident, later described as a "drunken misadventure," threw a spotlight on the new age vulnerabilities of one of the most protected buildings in the National Capital Region. The

---

[1] Michael D. Shear and Michael S. Schmidt, "White House Drone Crash Described as a U.S. Worker's Drunken Lark," *New York Times*, January 27, 2015, https://www.nytimes.com/2015/01/28/us/white-house-drone.html.

Secret Service never discloses the exact safety measures protecting the president's residence, but the layered system of defenses is thought to include spotters—human and electronic—and snipers and even coverage by anti-aircraft weapons.

The incident immediately prompted serious debate and recriminations, and sensitive discussions focused on the "what ifs?" of a drone carrying explosives or toxins right to the president's front door—and not at three a.m., when no officials or guests are expected outside on the White House grounds, but during a head-of-state arrival or Rose Garden news conference.

That lapse in White House security was up close and personal, but did not rock the markets. The stunning success of Iran's September 2019 drone and cruise missile attacks on Saudi oil facilities shocked war planners in Jerusalem and Washington and caused one of the largest spikes in crude prices in history. Those drone strikes proved that billions of dollars in military spending were useless against medium-tech, even low-tech, weapons that once were an American monopoly but now are fielded by dozens of countries and stateless militant groups. Readily available access to remote-controlled drone systems—with a tiny radar signature, that fly low and maneuver with precision—may have rendered many of the most prized air defense systems in the US and allied arsenals obsolete. These inexpensive drones, it is feared, could be used to take out infrastructure exponentially more valuable and impose additional costs in seeking new defenses.

And the Kremlin has learned the same lesson, as relatively inexpensive armed drones supplied to Ukrainian forces by Turkey and other nations, including the United States, were credited with blunting and, in some cases, even defeating columns of Russian tanks and armored personnel carriers following the unprovoked invasion in the spring of 2022.

"I'm amazed at how effective the TB-2 has been—the Turkish unmanned aerial system," Gen. James C. McConville, the United States Army chief of staff, told us.[2] "To me, unmanned aerial systems, or lethal drones, are for a lot of forces the most dangerous threat we're going to see moving in the future."

Ukrainian ground troops early in the war also began effectively using nonlethal drones outfitted with video cameras for surveillance, to spot and track Russian foot soldiers and vehicles on the streets of Ukrainian cities and villages, allowing the outgunned and outnumbered defenders to mount successful ambushes.

Their use is, without a doubt, going to spread. "We're going to see them, from violent extremists to Great Powers and all the way in between," McConville said.

In the battle of cost-effective lethality, armed drones are proving their value in Ukraine. The killer drones supplied to Ukraine cost from the tens of thousands of dollars to about $100,000. The Russian armored vehicles and tanks they destroyed cost from hundreds of thousands of dollars to a million or more.

But the Ukrainians also, and quickly, learned the limitations of their early fleet of off-the-shelf drones, and with tragic results. Commercial and hobby drones do not mask the signal between the pilot's remote position and the drone in flight. It did not take the Russians long to learn to track those radio signals—and rain lethal force down on the locations of the Ukrainian drone operators. Those combat casualties added to

[2] James McConville, "General James C. McConville, Chief of Staff of the Army" (transcript), Defense Writers Group, Project for Media & National Security, George Washington School of Media and Public Affairs, March 31, 2022, https://nationalsecuritymedia.gwu.edu/project/general-james-c-mcconville-chief-of-staff-of-the-army/.

the urgency of the United States and other nations supporting Ukraine to rush in military-grade drones that encrypt the signaling between operator and aircraft.

The concern also is much closer to home. On November 6, 2020, the Federal Aviation Administration put out a "Notice to Airmen" restricting flights over the Delaware home of Joseph R. Biden as he was on the cusp of winning the presidential election. Only law enforcement aircraft and planes from a nearby airport at Newcastle were exempted. The notice said that anyone flying a drone in the area could be subject to severe action from the Departments of Defense, Homeland Security, and Justice.

In the years after 9/11, the United States was first and best in adapting drones to become one of the signature tools of counterterrorism missions in Afghanistan, Iraq, Pakistan, and the Horn of Africa, carrying out both surveillance and strike missions. How did the United States not realize that its monopoly in drone technology, which proved its military and intelligence utility during two decades of Forever Wars, was not a lock forever? How did the United States fail to respond sooner to this proliferating threat? Is it too late?

One retired Air Force general, considered the godfather of drones at the Pentagon, offers a partial answer. David Deptula, who retired from the Air Force as a three-star general in 2010 after thirty-six years in uniform, was an early and forceful advocate of using Predator and Reaper drones for intelligence, surveillance, and reconnaissance—and attack.

"It's not like we were going to stop the proliferation of this technology," he told us.[3] "But guess what? Our US gov-

---

[3] David Deptula (retired Air Force general), interview with authors, September 17, 2021.

ernment policy actions made it worse." He cited a multilateral agreement, championed by the United States, called the Missile Technology Control Regime (MTCR), that was established in 1987 to limit the sales and distribution of missile technology and of missiles themselves, in particular, those capable of carrying weapons of mass destruction.

Deptula argues that drones simply are not missiles and should not be dropped into that category. "We just shot ourselves in the foot over a decade of preventing the US export of Predators and Reapers around the world," he said. "Instead of putting our arms around our friends and allies, and bringing them into our camp, we attempted to use this treaty to prohibit that. As a result, our adversaries capitalize on this, and begin selling their stuff to our friends and allies."

He said a galling example was when Jordan, one of Washington's staunchest allies in counterterrorism policy in the Arab world, joined Operation Inherent Resolve to halt ISIS in Syria. The Jordanians wanted American drones for gathering information and for strike operations. "We didn't sell them, claiming MTCR," Deptula said. "And so what did they do? They went to China. They open the door to our number one potential adversary to come in and infiltrate Jordan with all the other nefarious activities that the Chinese do. China sells them an inferior product, and instead of halting proliferation, the actions of the United States facilitated the actions of our adversaries in proliferating the advantages and technologies of drones around the world."

Drones had remained a niche capability through the 1990s, but interest grew across the military and intelligence leadership as attention began to be paid to the rise of Islamist terror groups. Although both the Clinton and the Bush administrations were criticized for not doing enough to detect and prevent the horrors of 9/11, there were high-level discussions in the summer of 2001, just before the terror attacks on the Twin Towers and the Pentagon, among senior administration, military, and intelligence officials about using an undisclosed base in Pakistan to target Osama bin Laden with drones should he be detected in hiding in Afghanistan.

It remains a little-discussed fact that Steven Hadley, who was serving as deputy national security adviser for President George W. Bush, convened a series of meetings in the summer of 2001 focused on using drones to hunt bin Laden. Around the table for these Deputies Meetings were the no. 2 officers across the national security establishment: Richard Armitage, the deputy secretary of state; Paul Wolfowitz, the deputy secretary of defense; Gen. Richard B. Myers, the vice chairman of the Joint Chiefs of Staff; John E. McLaughlin, the CIA deputy director, according to those briefed on the meeting.[4]

The change Hadley urged was to fly armed drones, capable of acting on the surveillance that the vehicle itself had collected, and in real time. CIA director George Tenet recalls that an unarmed Predator surveillance drone flying over Afghanistan on September 28, 2000, "observed a tall man in flowing white robes walking around surrounded by a security detail." Tenet wrote, "I don't know of any analyst who didn't subsequently conclude that we were looking at UBL."[5] The initials UBL often were used for Osama (Usama) bin Laden by some officials.

---

[4] Steven Hadley (former deputy national security adviser), interview with authors, September 21, 2022.

[5] George Tenet, *At the Center of the Storm* (New York: HarperCollins, 2007), 127.

Hadley sent a memo on July 11, 2001: Field armed Predators to search for bin Laden in Afghanistan, and be ready to strike, no later than September 1. Despite this order from the president's no. 2 national security adviser, a range of issues stood in the way: with interdepartmental funding, with legal approvals for a targeted killing, with integrating the Hellfire missile onto a drone designed for surveillance, and with having reliable, timely intelligence needed to target bin Laden and his associates.[6]

Ten days after Hadley's deadline: September 11, 2001.

After the 9/11 attacks, the government moved into overdrive, and the first armed Predator strike was carried out in October 2001, aimed at Mullah Omar, the one-eyed Taliban leader who had given safe harbor to bin Laden and al-Qaeda in Afghanistan. It missed. But in a foreshadowing of battle over turf and targets that sometimes complicated what should have been a team effort to fight terrorism, that strike was carried out by the CIA and was said by several officials to have completely surprised the senior generals carrying out the air war over Afghanistan. Drones would go on to play an unprecedented role in the wars that followed in Afghanistan and Iraq, and in countering terrorists across a range of nations where they sought refuge, including Pakistan and the Horn of Africa.

Today, adversaries are catching up and we don't have a clear plan for response. If a hostile drone carrying an explosive charge or an aerosol dispenser filled with a chemical or biological agent flew over a major sporting event in the United

---

[6] National Commission on Terrorist Attacks Upon the United States, *The 9/11 Commission Report* (Washington, DC: Government Printing Office, 2004), 211, https://www.govinfo.gov/content/pkg/GPO-911REPORT/pdf/GPO-911REPORT.pdf.

States, but one smaller, say, than the Super Bowl, it is not perfectly clear who has the authority and responsibility to spot the drone, track it, intercept it, and disable or destroy it.

The warnings exist. For years now, America's leaders have been provided with specific and verified information that first al-Qaeda and then ISIS have experimented with exactly such a threat.[7] Experts say it will just be a matter of time.

The ease of remotely piloted hobby vehicles interrupting civil and commercial life in the air has been proven innocently, if dangerously, when small drones have flown too close to Gatwick Airport, sowing fear, confusion, and shutdowns. At the height of the COVID pandemic, a socially distanced Major League Baseball game at Wrigley Field between the Chicago Cubs and Cleveland Indians was halted when a drone flew around the ballpark and even landed on the field. Federal Aviation Administration rules prohibit drones an hour before and after MLB games, so when this one landed—complete with flashing red lights—some players cowered in the dugout and others picked up bats to defend or attack. After seven minutes, during which time play was halted, the drone flew away.[8]

Yet, the rules on who will act, when they will act, and under what authority they will act are fuzzy in current regulations and still being sorted out.

The warning–action relationship on this threat is nowhere more of a priority than at US Northern Command and North American Aerospace Defense Command (NORAD), the com-

---

[7] Multiple interviews with senior national security and military officials, 2019–2022.

[8] Cole Little, "Watch: Drone Flying Above Wrigley Field Causes Delay," CubsHQ, September 16, 2020, https://www.cubshq.com/cubs-baseball/update/watch-drone-flying-above-wrigley-field-causes-delay-28872.

bined military headquarters in Colorado Springs. Northern Command was born in the bureaucratic reorganization after 9/11 to safeguard American territory and its airspace, and it has since become the government-wide 911 call center for hurricanes, floods, and wildfires, in addition to acts of terrorism.

Among the first to focus the United States on the drone threat to the homeland was General Lori Robinson, a four-star Air Force officer and the first woman to be promoted to leadership of a United States military combatant command. Robinson not only made history when she was sworn in as Northern Command commander in 2016, but her career is a heartfelt narrative of the challenges and lessons learned of the post-9/11 era—and, just as much, its tragedies and loss. Robinson's stepdaughter, Lt. Taryn Ashley Robinson, was an Air Force Academy graduate who died from injuries she received in a flight training accident in 2005.

After General Robinson was sworn in as the combined commander at Northern Command and NORAD, Robinson said of Ashley, "I knew she was peeking over the clouds, and I knew that she was saying, 'You go, Mom.' "

Robinson refocused NorthCom on the drone (and cruise missile) threat because she was deeply concerned that the United States had neither the technologies in place to down an enemy drone in real time nor the legal authorities in place for local, state, or federal law enforcement to do that in a real-time manner. She likes to ask, "Tell me, who is in charge here?"

NorthCom was among the significant reforms introduced to the Department of Defense and US military after 9/11. It was specifically designed to host an extensive array of interagency staff. And because NorthCom has to be poised for potential terrorist attacks emanating from anywhere in the world, veterans of the headquarters in Colorado Springs say its commanders receive intelligence reports that are among the broadest and the most detailed of the entire military. At NorthCom, military personnel work side by side with officials from the FBI, CIA, Homeland Security, the National Security Agency, and National Geospatial-Intelligence Agency—all of which are embedded at NorthCom. Northern Command also has specific infrastructure for sharing its warnings with state and local police departments and offers training and other support to those local agencies.

And drones are among its newest and most significant threats challenging the warning machine and action machine.

"Do you recall when the gyrocopter landed on the front lawn of the White House?" Robinson reminded us during an interview.[9] "So, the first was the gyrocopter. It was a chucklehead event. But you have to look no further than what's happening in the Middle East, and what they are doing with drones." For centuries, she said, this nation was protected by "a great big wall around the United States," the two oceans of defense that she refers to as America's "ponds, east and west." Now, Robinson noted, geography and distance are no longer sufficient to safeguard us from a drone attack.

"Ideas can transverse," she said. "And we have bad people here in the United States."

It did not take Robinson long to shake things up after taking over at NorthCom. During her first year on the job, the annual commander's exercise, called Ardent Sentry, planned the

[9] Lori Robinson (Air Force general), interview with authors, June 22, 2021.

response and recovery for a massive earthquake along the I-5 corridor in California, basically assuming Katrina-like catastrophe for the cities destroyed by shifting tectonic plates along the San Andreas fault.

In 2017, Ardent Sentry tried and tested for the first time a whole-of-government response—requiring bilateral cooperation with Canada—to a war game that posited a terrorist group simultaneously detonating improvised nuclear bombs in New York and Halifax.

For 2018, her final year in command, the potential flaws in the system responsible for warning and action—she politely calls it a "mismatch"—scrutinized by Robinson and her team were what she called "low radar cross-section threats," code for attack by ground-hugging cruise missiles from a nation-state adversary or by tiny, remotely operated drones sent by terrorists.

"There are a great many legitimate commercial and recreational opportunities for unmanned aerial systems operating in the domestic airspace system," Robinson said. "However, the potential exists for unmanned aircraft to be used by persons or groups with nefarious intent, especially given that their relative slow speed, low altitude, and small size make them capable of evading detection and tracking."

Robinson described for us the depth of her concerns about an undetected sneak attack by armed drone or cruise missile. "I slept with my cell phone on my chest," she said.

ISIS has "gone to school" on the use of drones by the American military in Iraq and Syria. Senior government officials tell us that they already have gathered evidence that the terrorist group has experimented with homemade drones carrying explosive charges or aerosol poisons.[10] Much of the information has been gathered from raids on ISIS laboratories or documents, hard drives, and thumb drives left behind when ISIS evacuates a town or village under allied assault. Even with the loss of its caliphate territory, this ISIS threat remains, either as individual ISIS cells in other countries or as "lone wolf" ISIS adherents who can turn to the internet and their local hobby shop to mount this threat.

The drone problem will only expand here at home, and quickly. As companies such as Amazon promise to fill the skies with drones making speedy deliveries of groceries, diapers, and electronics, the ability to discriminate friend from foe among these tiny unmanned devices in our nation's skies is an unexamined challenge. According to the Congressional Research Service, the Federal Aviation Administration had registered about 1.7 million unmanned aerial vehicles as of September 2020. That number is estimated to grow to 2.3 million by 2024, with 1.5 million recreational users and about 800,000 commercial unmanned aerial vehicles.[11]

In early 2018, officials told us privately that a drone attack on a political leader, large public sports event, or outdoor concert was simply a matter of time. Then, in August 2018, Venezuelan leader Nicolas Maduro was attacked by two drones armed with explosives at an outdoor speech in Caracas. Drones carrying explosives also attacked the home of the Iraqi prime minister in November 2021. Iraqi government security forces were able to down two of the three drones making the

---

[10] Lori Robinson, interview with authors, June 22, 2021.

[11] Congressional Research Service, "Protecting Against Rogue Drones," *In Focus* IF11550, version 3, updated, September 3, 2020, https://crsreports.congress.gov/product/pdf/IF/IF11550.

attack; the third wounded six security guards and damaged the residence.[12]

Here in the United States, officials say, clear rules and lines of authority are not yet in place for deciding who must monitor and who should bring down a suspicious drone.

According to officials who have focused on the drone threat, a significant effort to clarify authorities and describe enforcement mechanisms only came late, in 2018, when Congress passed the Preventing Emerging Threats Act.[13] The bill ordered the Department of Homeland Security and the Department of Justice "to authorize their personnel to act to mitigate the threat that unmanned aircraft (i.e., drones) poses to the safety or security of facilities or assets, through a risk-based assessment."

Specifically, the legislation pinned onto DHS the authority to

- Detect, identify, monitor, and track the drone, without prior consent
- Warn the drone's operator
- Disrupt control of the drone, without prior consent
- Seize or exercise control of the drone
- Confiscate the drone
- Use reasonable force to disable, damage, or destroy the drone

---

[12] John Davison and Ahmed Rasheed, "Iraqi PM Decries 'Cowardly' Attack on his Home by Drones Carrying Explosives," Reuters, November 8, 2021, https://www.reuters.com/world/middle-east/iraqi-pm-chairs-security-meeting-after-drone-attack-residence-2021-11-07/?utm_source=newsletter&utm_medium=email&utm_campaign=newsletter_axiosam&stream=top.

[13] Preventing Emerging Threats Act of 2018, S.2836, 115th Cong. (2018), https://www.congress.gov/bill/115th-congress/senate-bill/2836.

"Any drone seized by DHS or DOJ is subject to forfeiture to the United States," the legislation said, adding: "DHS shall: (1) evaluate the threat from drones to U.S. critical infrastructure and to domestic large hub airports; and (2) assess the threat of vehicular terrorism and its activities to support emergency response providers and the private sector to prevent, mitigate, and respond to vehicular terrorism."

Clear, right? Not to those specifically charged with creating a new organization within DHS to do that. A senior official involved in the effort described the scrambling to put the new law into action and the questions it brought. "How do you identify friend from foe?" the official said. "How do you track the unpiloted vehicle carrying a kidney transplant versus one that might be carrying ordnance or some other nefarious material?" Not easy.

The legislative effort began by stitching together a series of logical if ad hoc drone protection procedures that were already in place. The Defense Department has the authority to defend all of its bases, forts, fields, and facilities from drones. The National Capital Region is a protected area. Major airports, as well. Anywhere the president travels. Large public events, like the Super Bowl. Significant gatherings, such as the United Nations General Assembly. And specific counter-drone protocols have been set in place for occasional but sensitive missions, such as when the US Marshals Service transports a fugitive by road.[14]

An example of these defenses designed around specific, valuable locations surfaced in August 2021, when the nuclear laboratory at Los Alamos, home of the first atomic bomb,

---

[14] Interviews with multiple senior national security officials, who spoke on condition of anonymity, Summer 2022.

reported a worrisome number of unauthorized drone flights over its restricted air space.[15] "We can detect and track a UAS [unmanned aircraft system], and if it poses a threat, we have the ability to disrupt control of the system, seize or exercise control, confiscate or use reasonable force to disable, damage or destroy the UAS," Unica Viramontes, senior director of lab security, told the Associated Press. No details of the defensive systems were disclosed.

Although there are cases when the solution is bringing down the drone with bullets, missiles, netting, or even lasers, far more effective and efficient are electronic measures, such as jamming the guidance system or frying the circuits. "A lot of that technology is still in that gray area of getting the FAA and the airlines comfortable that it is not going to impact air traffic," said one official. Nobody wants the beams that bring down a hostile drone to also cause a civilian airliner to go off course or crash.

But experts on the science and technology side of countering drones caution that these systems can identify a hostile drone about 50 percent of the time, and that a target, once identified, can be taken down about 50 percent of the time—which does not yield high confidence. And the loss of public confidence in a high-profile attack could be catastrophic, even if there was not a catastrophic loss of life.

"Say, you had a stadium full of a hundred thousand people at Ann Arbor," one official mused. "If somebody were to just drop, you know, powdered sugar, the amount of impact that that

---

[15] Associated Press, "So You Want to Fly a Drone over a Nuclear Weapons Lab…" *Defense News*, August 24, 2021, https://www.defensenews.com/unmanned/2021/08/24/so-you-want-to-fly-a-drone-over-a-nuclear-weapons-lab/.

would have would be catastrophic. It's never about, 'Will they drop a biohazard?' Just the fact that a drone came and dropped anything, the confidence in our ability, as a nation, to protect our citizens goes out the window. And the economic loss."

There is no national network of drone detection radars or sensors yet in place, and so the risk exists equally, if not more so, to critical infrastructure nationwide, including electricity systems, gas pipelines, and water systems, the official said. But those systems also bring an unexpected benefit, at least to tracking drones. Most large public utilities have extensive security systems and sensing monitors, as do oil and gas facilities and their pipelines, as do airports. As do major port cities. And some metropolitan areas, including Washington, DC, with so many government offices, and New York City, a financial center that also hosts the United Nations, have extensive systems of radars monitoring the airspace, not just because of what is located there but also because they have multiple airports located near dense population zones. DHS already is at work on plans to stitch together these separate systems into an initial patchwork grid to share information about possible drone activity across the nation.

Even officials deeply involved in the counter-drone enterprise acknowledge a major hurdle is protecting the nation while preserving civil liberties. To use electronic warfare to down a drone requires taking over or canceling the signal, and that gets into issues of privacy.

"In order for me to take control of that drone, or to disrupt that drone, I have to disrupt the signal, whether it be a radio frequency signal or a Wi-Fi signal," one official explained.[16] "I

[16] Interview with national security officials, conducted on rules of nonattribution, Summer 2022.

have to take it. I have to disrupt it. I have to grab it. And especially when you take control of the drone, you're grabbing that signal, decoding it."

To explain the civil liberties challenge, the official used the analogy of a letter delivered by the US Postal Service. "Some of the technology just sees the envelope and says, 'Yep, there's a letter,'" the official said. "Some of the technology opens up that letter, opens up the envelope, and pulls the letter out. And some of it actually reads that letter." Those counter-drone systems that treat the control signal in the equivalent way of opening a letter and reading it could violate wiretapping laws and rights to privacy, without proper authorities and legislation being in place.

Those sensitivities have driven some research and development into drone countermeasures to be conducted outside the United States to avoid colliding with civil liberties issues even before a new system is deployed. Officials say that Israel is among those nations that have hosted these technological testing efforts, and enthusiastically.

"We don't live in an environment where a drone is more likely than not weaponized," the official told us. "I mean, that's the reality of Israel."

Among new proposals would be to require drone owners to register and add a tail number to their drone, much like those on larger aircraft, as well as specifically limiting the size and weight of a drone that could fly over a populated area—while prohibiting all flights over "open-air assemblies," such as sporting events, political rallies, and outdoor concerts and fairs.

What still remained was a hodge-podge of "Who is in charge here?" Local police? State authorities? Only the FAA or another federal agency? Northern Command? In short, the warning and action machines have yet to be brought into sync. And some of the machinery is yet to be built.

Hoping to illustrate how it had codified rules of the sky for remotely piloted vehicles, the Federal Aviation Administration in September 2020 released its "Drone Response Playbook for Public Safety."[17]

"We are at an exciting time in aviation, where drones are being safely integrated into our national airspace for recreational, commercial, and public safety uses," the agency said, in language that sounded more like a public service announcement than a national security solution. That concern was saved for the next sentence: "However, unauthorized operations can cause potential hazards to people and property both in the air and on the ground."

The goal of the document, the FAA said, was to "help determine the difference between authorized and unauthorized drone operations and what actions public safety agencies may take." The document also noted the overlapping jurisdictions and authorities, noting that users of the FAA manual should study "local rules and regulations."

For clarity, the manual listed a primary set of restrictions:

- Drone flights within 3.45 miles of a qualifying event at a stadium or sporting venue without an FAA authorization
- Flights over people without an FAA waiver
- Night operations without an FAA waiver

---

[17] Federal Aviation Administration, *Drone Response Playbook for Public Safety* (Washington, DC: Federal Aviation Administration, 2020), https://www.faa.gov/uas/public_safety_gov/public_safety_toolkit/media/Public_Safety_Drone_Playbook.pdf.

- Failure to give right-of-way to manned aircraft without an FAA waiver
- Operations beyond visual line of sight without an FAA waiver

Then it got down to serious business, listing these prohibited activities:

- Operation while under the influence of alcohol and/or drugs
- Hazardous and/or unsafe operations
- Carrying illegal narcotics
- Carrying hazardous materials
- Operation of a drone that is equipped or armed with a dangerous weapon

An acknowledgment of the increasing risks from drones is not to say that enduring threats from long-range cruise missiles or from a new generation of hypersonic missiles should be relegated to second tier. Those fast, hard-to-track missiles can carry weapons of mass destruction and conventional warheads with such accuracy as to vastly increase their lethality. But those technically advanced weapons are only fielded by advanced powers, and the threat is recognized. Drones are becoming a low-cost cruise missile for lower-tier powers, non-state actors—and individuals with a radical agenda, left or right.

The four-star officer serving in the top position at Northern Command and North American Aerospace Defense Command in 2022 said he spends much of his time focused on that high-end threat of cruise missiles and hypersonic weapons. But the commander, Gen. Glen D. VanHerck, also told us that the threat from drones "will only continue to grow—and it's just a matter of time before we see another incident where we have to respond."[18]

General VanHerck did note that drones weighing less than fifty-six pounds, the most common in the skies today, are not his mission but are a law enforcement responsibility nationwide.

He said the agenda for dealing with the drone threat that falls under his responsibility includes stitching together current surveillance, radar, and other tracking data to create a fuller picture of the skies, as well as the ability to deeply interrogate that information to distinguish, say, a bird from a drone—and then manage at network speed, with artificial intelligence and machine learning, to make the right decision, and in time.

A key element is the Pathfinder Program, which increases the ability to gather, organize, and assess bulk radar and other tracking data—because, on most days, only 2 percent of that available data can be or is used, he said. "We're taking the raw data, 100 percent of the information, and fusing that, and using artificial intelligence and machine learning, and distributing that information to gain time and space, if you will," he said.

Remember the 2015 "chucklehead" incident when a drunk lost control of a hobby gyrocopter, which landed on the White House lawn, which set off a Secret Service emergency scramble?

Afterward, the military used the Pathfinder Program to retrospectively analyze the data from the incident. "When you took

---

[18] Gen. Glen VanHerck, "General Glen D. VanHerck, Commander, United States Northern Command and North American Aerospace Defense Command" (transcript), Defense Writers Group, Project for Media & National Security, George Washington School of Media and Public Affairs, April 25, 2022, https://nationalsecuritymedia.gwu.edu/project/general-glen-d-vanherck-commander-united-states-northern-command-and-north-american-aerospace-defense-command-2/.

that data and you fused all available data from all sources, you were actually able to see the full radar track," he said. "Where if we didn't do that [crunch the data through Pathfinder], no radar track saw it. So I'm talking military information, commercial information, FAA information, multiple [sources of] information. That's the idea behind Pathfinder, taking the raw data and fusing it from multiple sources and allowing it to be processed and disseminated."

In other words, if the right system had been in place—the right "machine"—the gyrocopter would not have snuck up on anyone. It would not have been a surprise because it could have been tracked in real time.

Authorities are clearer for the US military abroad. American military commanders in conflict zones overseas and whose forces routinely come under attack from hostile drones have clear authority to take any steps required to protect their forces, whether on land or at sea. A Navy amphibious warship, the USS *Boxer*, entering the Strait of Hormuz in the summer of 2019, faced a suspected Iranian drone approaching on a hostile flight path. The ship downed the drone.[19] Officials did not disclose which weapons were used, but the ship carries a variety of short-range missiles and rapid-fire guns, as well as sophisticated electronic jamming systems. (Officials briefed on the incident said that the ship's commander opted for a less "kinetic" option, ordering jamming rather than missile or gunfire.[20]) Similarly, American bases in conflict zones are protected by a

[19] Sam LaGrone, "Updated: USS *Boxer* Downs Iranian Drone in 'Defensive Action,'" USNI News, July 18, 2019, https://news.usni.org/2019/07/18/uss-boxer-downs-iranian-drone-in-defensive-action.

[20] Interview with Pentagon civilian and military officials, conducted under rules of nonattribution, Summer 2022.

variety of radar systems and anti-aircraft weapons, but drones provide a difficult target because they are small, offering a low radar cross section.

Before retiring as commander of American forces in the Middle East in 2022, Gen. Kenneth F. McKenzie visited the United Arab Emirates to discuss the increasing threat of drones, both to military forces and the Gulf's critical oil infrastructure. In an interview with Emirates News Agency, he described advances in technology to knock drones out of flight—but he emphasized the need to halt drone attacks before they are even launched.

"We are working with our partners here in the region and with the industry back in the United States to develop solutions that would work against drones," McKenzie said.[21] "We would like to work against drones in what we call 'Left of Launch,' "—per military jargon, before they can be launched. "And if you can't do that, you will certainly be able to shoot them down as they reach their intended target."

Accelerated programs to develop technologies to prevent a drone attack now underway include ground-based systems to fire projectiles or nets, as well as low-level jamming that would affect a drone's control function but not that of other aircraft in the vicinity. Some of these tactical jammers are small enough to be operated by an individual soldier. Officials say that as adversaries learn to harden their drones against jamming, defenses likely will rely on microwaves or lasers, as well

---

[21] Binsal Abdulkader, "US to Help UAE Improve Air Defence System, Stop Drones Before Launching: CENTCOM Commander," Emirates News Agency, August 2, 2022, https://wam.ae/en/details/1395303018778.

as other directed-energy weapons, to burn the operating systems and disrupt connections to the remote pilot.[22]

A review of Pentagon procurement notices shows the vast amounts being spent to develop and field systems to defend against drones. In 2021, the Army budgeted $190 million to counter drones; the Navy, $73 million; the Marine Corps, $38 million; the Air Force, $29 million; and, in a telling separate budget line, the military's Special Operations Command—which fields Green Berets, Navy SEALs, and the Army's Delta Force—had its own $38 million budget for systems to defend against drones.[23] Those numbers are only growing.

The contracts have gone to some of the nation's largest defense contractors and to some smaller, niche enterprises. The systems are designed to smack into a hostile drone or fry the components or scramble control signals of an individual drone and even swarms of remotely piloted vehicles. These new defenses have names that sound borrowed from Marvel Comics, including ANVIL, Self-Protect High-Energy Laser Demonstrator (SHiELD), and Tactical High-Power Operational Responder (THOR).

However, even those proving their capability are not yet deployed in sufficient numbers—at least to protect the entire homeland as opposed to deployed American forces in a specific location. As recently as 2020, the Department of Homeland

---

[22] James Marson and Stephen Kalin, "The Military's New Challenge: Defeating Cheap Hobbyist Drones," *Wall Street Journal*, January 5, 2022, https://www.wsj.com/articles/the-militarys-new-challenge-defeating-cheap-hobbyist-drones-11641401270.

[23] Jon Harper, "Pentagon Gets $7.5 Billion for Unmanned Systems," *National Defense*, May 27, 2021, https://www.nationaldefensemagazine.org/articles/2021/5/27/pentagon-gets-$7-5-billion-for-unmanned-systems.

Security's inspector general found that DHS still had only "limited capabilities to counter illicit unmanned aircraft systems."[24]

Even some systems to counter drones that could prove useful in a conflict zone are not compatible for use in the United States, given the ubiquitous presence of private and commercial civil aviation whose communications and guidance systems, obviously, cannot be disrupted, even in defense against a hostile drone. Of course, employing lasers in densely populated areas won't be an effective option either.

By early 2022, the Department of Homeland Security was touting the development of its "novel and innovative technologies" to defend against unmanned aerial systems. In particular, work was focused on the UAS Traffic Management (UTM) and Air Domain Awareness (ADA) systems. The programs "will detect UAS in the air from the ground to 500 feet up, identify the target and its attributes in order to determine the threat level, and if necessary, bring the UAS down safely," the department said. "That is, take it down without causing collateral damage to people or property on the ground. UTM will establish airspace flight corridors, geo-fencing, route planning, terrain avoidance guidance, and weather alerts, among other capabilities."[25]

But if, as experts warn and officials agree, a drone attack on the United States is inevitable, when it happens, will it be easy to point the finger at US government officials who should have seen it coming? Or are US policymakers wise to

---

[24] Office of Inspector General, *DHS Has Limited Capabilities to Counter Illicit Unmanned Aircraft Systems* (Washington, DC: Department of Homeland Security, June 25, 2020), https://www.oig.dhs.gov/sites/default/files/assets/2020-06/OIG-20-43-Jun20.pdf.

[25] "Controlling Our Airspace in the Age of Drones," Department of Homeland Security, Science and Technology, last updated January 7, 2022, https://www.dhs.gov/science-and-technology/controlling-our-airspace-age-drones.

avoid imposing regulations and procedures that may be moot the minute they are enacted because technology evolves and the hazards and appropriate responses come into better focus?

This actually may be a case where government attention, however belated, deserves credit for trying to fix, update, and upgrade the warning machine and the action machine, which have been struggling to get ahead of the curve on this rapidly transforming technological challenge.

In late April 2022, the Biden administration unveiled the nation's first "Domestic Counter-Unmanned Aircraft Systems National Action Plan,"[26] which provided eight specific lines of effort to protect the nation from dangers posed by remotely piloted vehicles:

1. Work with Congress to enact a new legislative proposal to expand the set of tools and actors who can protect against UAS by reauthorizing and expanding existing counter-UAS authorities for the Departments of Homeland Security, Justice, Defense, State, as well as the Central Intelligence Agency and NASA in limited situations. The proposal also seeks to expand UAS detection authorities for state, local, territorial and Tribal (SLTT) law enforcement agencies and critical infrastructure owners and operators. The proposal would also create a Federally-sponsored pilot program for selected SLTT law enforcement agency participants to perform UAS mitigation activities and permit critical infrastructure

[26] "Fact Sheet: The Domestic Counter-Unmanned Aircraft Systems National Action Plan," White House, April 25, 2022, https://www.whitehouse.gov/briefing-room/statements-releases/2022/04/25/fact-sheet-the-domestic-counter-unmanned-aircraft-systems-national-action-plan/.

owners and operators to purchase authorized equipment to be used by appropriate Federal or SLTT law enforcement agencies to protect their facilities;

2. Establish a list of U.S. Government authorized detection equipment, approved by Federal security and regulatory agencies, to guide authorized entities in purchasing UAS detection systems in order to avoid the risks of inadvertent disruption to airspace or the communications spectrum;
3. Establish oversight and enablement mechanisms to support critical infrastructure owners and operators in purchasing counter-UAS equipment for use by authorized Federal entities or SLTT law enforcement agencies;
4. Establish a National Counter-UAS Training Center to increase training accessibility and promote interagency cross-training and collaboration;
5. Create a Federal UAS incident tracking database as a government-wide repository for departments and agencies to have a better understanding of the overall domestic threat;
6. Establish a mechanism to coordinate research, development, testing, and evaluation on UAS detection and mitigation technology across the Federal government;
7. Work with Congress to enact a comprehensive criminal statute that sets clear standards for legal and illegal uses, closes loopholes in existing Federal law, and establishes adequate penalties to deter the most serious UAS-related crimes; and
8. Enhance cooperation with the international community on counter-UAS technologies, as well as the systems designed to defeat them.

The single most important aspect of the action plan and accompanying legislative proposals, according to a Biden administration official who was on the interagency team that developed the action plan, is that the current legislation granting authorities for defending against drones, passed in 2018, is set to expire before the end of 2022.[27] The official said that this ticking clock "added to a sense of urgency specifically with respect to our legislative proposal, which would renew some of those expiring authorities and then add additional ones that we think are necessary to keep pace with the threat. That's one additional thing that moved us to act." New legislation, called "Safeguarding the Homeland from the Threats Posed by Unmanned Aircraft Systems Act of 2022," was introduced in August 2022 by a bipartisan group of senators.[28] No final action had been taken by late 2022. The bill stalled.

Although funding totals are not included in the proposed legislation, and would be a separate budget negotiation between the president and Congress, the question of increasing resources to counter the threat is obvious. According to the Biden administration official, the Departments of Homeland Security and Justice have provided federal assets for drone protection to less than 1 percent of the more than one hundred thousand public events that state and local authorities have specifically asked for a "Special Event Assessment Rating." These include major professional sporting events and large holiday celebrations in major metropolitan areas, including New Year's

---

[27] Author interview with senior administration official, conducted on background attribution, Summer 2022.

[28] Safeguarding the Homeland from the Threats Posed by Unmanned Aircraft Systems Act of 2022, S.4687, 117th Cong. (2022), https://www.congress.gov/bill/117th-congress/senate-bill/4687?s=1&r=3.

Eve and July Fourth. "If the provisions that currently are on the books were to expire, that goes down to zero percent," the official said. If Congress lets the current authorities expire without at least renewing them, it would mean that future Super Bowls and Indy 500 races that should be no-brainers for federal protection would become "Sorry. No."

The official acknowledged the complicated questions regarding authorities to track and, if necessary, bring down a drone, especially given privacy. That remains uppermost in the minds of some members of Congress and civil liberties groups, who are urging the administration to go slow in expanding these authorities. Thus, any new legislation must be written to not only expand defenses against drones but also clarify who is allowed to do what, making the action and warning machines operate more effectively and within the law. Channeling an inner Johnny Cash, the official said the Biden administration is attempting to "walk the line that gets to the art of the possible in terms of legislation, which is no easy feat these days, especially on complicated issues of technology and privacy and civil liberties."

The central question remains: Who is in charge?

Perhaps the best way to answer is by comparing the threat of drones to other significant security threats, which are mitigated by a whole-of-government approach, with no single, individual czar managing the effort.

"What if I said, 'Who was in charge of counterproliferation?' " the official said. "My answer would be, well, there's an intelligence community that detects proliferators. There's a law enforcement community that charges those who break US law. There's a State Department community that rallies partners to do their own part in this effort. There's a mitigation community that works for critical infrastructure operators to protect against that which does proliferate. But there is no one proliferation lead or counterterrorism lead or for many, many other multifaceted national security problems."

Perhaps it makes sense in a multilayered threat environment presented by drones—where there are different parts of the federal government and there are state and local law enforcement agencies, and critical infrastructure owners and operators, and other private-sector contractors—that the best system is a mesh of shared goals and shared capabilities.

But the gaps remain painfully, dangerously obvious. On July 21, 2022, just days before we spoke with the senior official, Washington, DC's, Reagan National Airport, within sight of the Capitol dome, was shut down for about half an hour when a drone was spotted near a main runway. After spotting the drone, scores of flights were immediately delayed or rerouted. Airport officials notified the Federal Aviation Administration. The FAA then notified the Metropolitan Police Department. Queries about the delays were referred to the individual airlines.[29] Samantha Vinograd, the acting assistant secretary of homeland security, said that the Transportation Security Administration had tabulated about two thousand drone incidents near domestic airports since 2021—and that evasive action was required by pilots in sixty-five of those events, when the drone came dangerously close.[30]

---

[29] Katherine Shaver, "Drone Sighting Briefly Stops Air Traffic at Reagan National," *Washington Post*, June 21, 2022, https://www.washingtonpost.com/transportation/2022/07/21/reagan-national-delays-drone/.

[30] Tanya Snyder, "Errant Drone Briefly Shuts Down D.C. Airport," *Politico*, July 21, 2022, https://www.politico.com/news/2022/07/21/drone-dc-airport-shutdown-00047213.

Josh Geltzer, deputy assistant to the president and deputy homeland security adviser, described what prompted the Biden administration's sense of urgency in working to fix the government machines that focus on the drone threat.

"When I think about the threat to the homeland in particular, I think about it in two categories," Geltzer said in an email exchange.[31] "One is what we have already seen in the homeland. And what we have already seen includes activity at airports that could threaten incoming planes. Thankfully that has not yielded any awful incidents. But that's thanks to evasive maneuvers and quite costly pauses in operations."

The second category, he said, "is things we've seen happen abroad. We have to think it's only a matter of time until they happen here, or at least are attempted here. And in that category, I'd put even worse activity, including outright arming of drones. They've been used in assassination attempts.

"If we use what happens abroad as a precursor of what could soon happen here," he concluded, "I think we see an even more challenging array of threats on the horizon."

---

[31] Josh Geltzer (deputy assistant to the president and deputy homeland security adviser), email exchange with authors, August 22, 2022.

CHAPTER 8

STORMS

Everything, Everywhere, All at Once

The military mission was traditional, straightforward, right from the manual: Navy warships would ferry Marines, their amphibious landing craft, and assault jets to the Western Pacific for a landing on a hostile island shoreline. Assigned to the assault were an Amphibious Ready Group consisting of an amphibious assault ship—basically, a small aircraft carrier for launching short take-off and landing jump-jets as well as helicopters—a transport dock ship for landing Marines on contested beaches via landing boats or air-cushion vehicles, and support vessels. Combat personnel assigned to the mission included a Marine Expeditionary Group of an infantry battalion, usually about twelve hundred Marines, bolstered by artillery and armored vehicles, communications systems, and combat logistics to sustain the operation.

In advance of the mission, Navy meteorologists carefully tracked an autumn Pacific storm as the twirling winds came together—still at a safe distance. But by the time the winds reached catastrophic typhoon level, the storm had changed course, and the superstorm slammed into the Navy and Marine forces at sea. Giant troughs scattered the warships out of their formations. Howling winds made air operations and air rescue impossible. Communications were shredded.

The cascading effects of the extreme storm only grew worse for the failing operation because years of climate change meant that local islands and the local populations, still recovering from previous mudslides, power failures, and broad infrastructure disruptions brought by other typhoons that season, were of little assistance. There was no safe port in this storm.

This mission occurred seven years in the future, in October 2030, as played out in the first-ever war game conducted by the Navy and Marine Corps to assess the challenge that climate change is presenting to the military's ability to carry out its mission. The table-top exercise, held in late June 2022, garnered scant public attention, but it sounded a clarion across the maritime services.

Although elected leaders, TV talking heads, and many in the public still argue the reality of climate change, and its politics, the military does not have that luxury. Climate change is a reality adding a powerful destabilizing force to already fragile, unstable areas of the world. Once-in-a-century ocean storms happen several times each season. Drought prompts food shortages, civil unrest, mass migration. Island nations that once served as idyllic homelands are vanishing under rising seas. All these make the Defense Department's efforts to combat global instability that much harder, even as DOD has to admit that the American armed services are the world's largest consumers of fossil fuels, contributing to the greenhouse effect and global warming.

The enemy always gets a vote, or so says a military axiom, and climate change is a new adversary.

"We are looking at the impacts of climate change because it makes us better warfighters," said Meredith A. Berger, the assistant secretary of the Navy for energy, installations, and environment, who organized the climate change exercise. "The Navy and Marine Corps must address climate change in our readiness and operations in order to maintain every advantage to fight and win."[1]

It now is clear in retrospect that the threat of terrorism, even in the wake of the 9/11 attacks, was never an existential threat; even on its best day, al-Qaeda never truly threatened the existence of the United States. But climate change is, indeed, an existential threat, Berger said. "This is something that we are hearing from the president all the way down to sailors and Marines," she said. "We're hearing this from partners, too. This is something that everyone is calling an existential threat, and we are reacting in kind."[2]

The climate war game played out in parallel to a larger disaster movie playing out across America and the world. It's also a reality show.

Catastrophic wildfires have broken out from the hot summer season, sparking earlier and burning later. Military bases in the heartland find their runways unusable—not cratered by

---

[1] "The Department of the Navy Hosts Climate Tabletop Exercise" (press release), America's Navy, June 29, 2022, https://www.navy.mil/Press-Office/Press-Releases/display-pressreleases/Article/3079453/the-department-of-the-navy-hosts-climate-tabletop-exercise/.

[2] Meredith A. Berger, "Assistant Secretary of the Navy Meredith A. Berger, Energy, Installations, and Environment" (transcript), Defense Writers Group, Project for Media & National Security, George Washington School of Media and Public Affairs, June 21, 2022, https://nationalsecuritymedia.gwu.edu/project/assistant-secretary-of-the-navy-meredith-a-berger-energy-installations-and-environment/.

enemy ordnance but covered by floodwaters. Major bases on the Atlantic and Pacific coasts—gargantuan Navy and Marine Corps installations critical to American defense and force projection—confront the near-certain threat of being inundated by rising sea levels. As the air gets hotter and wetter, warplanes and helicopters have a difficult time taking off, which requires them to burn even more fossil fuels for lift. More periods during the year will be far above the heat and humidity levels that require drill sergeants to declare a Black Flag Day, when all training and exercise not related to a real-world, operational mission are canceled; critical military operations on those days, at home and around the globe, will increase risks to every soldier out in the sun. Melting ice caps could release horrific pathogens that have been frozen in suspended animation for ages—and the Navy would be first to sail into, or stumble upon, them as its warships explore regions of the Arctic newly accessible by retreating polar ice.

While the Defense Department came late to acknowledging and acting on risks posed by climate change to its installations and operations, experts note that the armed forces have come around and are on the hunt for solutions. Effects from climate change that directly impact the ability of the armed forces to do what they do are challenges the military should be prepared to assess and solve because these activities are core to the mission. The Pentagon is all about mission-first and is going to figure it out because it has no other choice.

It will take money, which has not yet been sufficiently earmarked. It will take attention, which is finally refocusing on this risk. But it can be done.

"What I think are harder problems to understand are some of the second- and third-order risks from climate change that are shaping the security landscape," said Erin Sikorsky, a retired CIA analyst who focused on environmental risks during a decade-plus career in government.[3] "It's not just a military problem by any stretch. It's all the arms of a foreign policy. That's where I think a lot of attention is needed still. How do we build institutions and the machinery to deal with that reshaped landscape?"

Early in her CIA career, in the years before and then just after 9/11, she led teams looking at conflict and instability in Africa and the Middle East. Even though the focus was war between states, civil strife within their borders, and terrorism, she described a eureka moment of coming to understand an underappreciated, destabilizing power that was fueling those threats.

"I realized that we were parsing the drivers of those things, and that environmental and climate issues were coming up time and again, and we didn't, really, on our teams, have the tools or capacity to really understand that," Sikorsky said. "That wasn't part of the traditional suite of things that the analysts looked at. It just started me thinking more and more about the security implications of climate change, and what those look like and what kind of tools and approaches we needed in the intelligence community, particularly—but also in national security, writ large, to deal with that risk."

There is little disagreement that climate change is real, and is here, but the US government—if belatedly—has mostly approached the crisis as a problem to be solved with a series of negotiations on treaties, like the Kyoto and Paris agreements, or by pressing for fuel efficiencies and emissions standards, and

---

[3] Erin Sikorsky (retired CIA analyst), interview with authors, April 18, 2022.

seeking alternative energy resources. But a policy that relies on mitigation is unlikely to go fast enough or far enough.

As Sikorsky was promoted up the analytical side of the CIA house, she kept demanding attention to the climate change impact on security, an effort that led her to be named deputy director of the Strategic Futures Group on the National Intelligence Council and tasked with coauthoring the quadrennial Global Trends report and the intelligence community's environmental and climate security analyses.

Examples? In advance of the brutal civil war in Syria, drought played a key role in prompting migration to urban areas, which put pressure on the Assad government, whose political institutions were overwhelmed. This only exacerbated latent tensions between ruling Alawites, a minority, and the majority Sunnis in the Syrian population.

In recent years, Iran and Iraq and Lebanon have suffered street protests over high energy prices and lack of water. Although climate change was not the only driver—those nations are wracked by poor governance, corrupt political institutions, and feeble infrastructure—rising global temperatures have worsened those shortages and prompted migration from rural to urban areas, where local services simply cannot provide sufficient food and water. Unrest and violence between tribes and religious denominations are the outcome. The same drivers are at play in the conflict in Ethiopia, Africa's second-most-populous country, over the Tigray region.

World Bank analysis out to 2050 describes a horrific rise in climate-related migration, with these projections: East Asia and the Pacific, 49 million people. South Asia, 40 million. North Africa, 19 million. Latin America, 17 million. Eastern Europe and Central Asia, 5 million. By far the most catastrophic situation is in Sub-Saharan Africa, with 86 million climate-related migrants predicted by 2050.[4]

As commander of US Africa Command, Army Gen. Stephen J. Townsend is responsible for the military's role in defending American interests on the continent, which mostly means working to improve the professionalism of partner state armed forces. He told us that the threat of climate change ranks as his no. 3 challenge in carrying out the mission, after malign Russian influence and regional terrorism. "Climate change and the environment: Whether someone is a believer in climate change or not, the environment is definitely affecting lives in Africa," he said. "Drought, famine, desertification—all of these things I think will continue to be challenges that are just tough to accomplish."[5]

Perhaps there is no more dramatic example than in Somalia, which has the longest shoreline of any country in Africa. Historically, the fish stocks in waters off the Horn of Africa provided life and livelihoods. But in recent decades, especially following the civil war in Somalia, overfishing in close-in territory waters—much of it illegal, by foreign factory fishing boats—drastically depleted the fish supply. Local fisherman could not compete with the technologically more advanced

---

[4] Juergen Voegele, "Millions on the Move: What Climate Change Could Mean for Internal Migration," Voices (blog), World Bank, November 1, 2021, https://blogs.worldbank.org/voices/millions-move-what-climate-change-could-mean-internal-migration#:~:text=Taken%20together%2C%20projections%20across%20all,and%20Central%20Asia%2C%205%20million.

[5] Stephen J. Townsend, "General Stephen J. Townsend, Commander, US Africa Command" (transcript), Defense Writers Group, Project for Media & National Security, George Washington School of Media and Public Affairs, July 28, 2022, https://nationalsecuritymedia.gwu.edu/project/general-stephen-j-townsend-commander-us-africa-command/.

foreign fishing vessels that were robbing them of sustenance and a source of income. Their traditional way of life ended. The result? Somali fisherman had boats but no fish, so they turned to piracy, hijacking foreign vessels for ransom, the most famous being the case of Captain Richard Phillips of the *Maersk Alabama*, who was rescued by Navy SEALs.

Although the pirates and terrorists in Somalia may share little ideology, US officials say there is a dangerous convergence of interests. Al-Shabaab, American officials say, supports pirate crews with weapons and logistical help, in exchange for a cut of the ransom to support its terrorist activities in the region.

"It's an added factor that stresses and strains governments, creates instability, creates fragility and risk of conflict," Sikorsky said. And the risks are not contained within those borders. "It also creates geopolitical risks over what countries will compete over."

After retiring from the CIA in 2020, Sikorski served as the founding chair of the Climate Security Advisory Council, created by Congress with a mandate to enhance collaboration and communication between the intelligence community and government scientific agencies. She now is director of the Center for Climate and Security, a policy institute.

She said that the government action machine clearly heard the ominous assessments on climate from the warning machine. "People always agreed," she said. "They'd say, 'Yeah, I think you're right.' But then when it came to actually taking action, and putting resources toward it, then, no. They'd say, 'It's not a security problem. It's not an intelligence community problem.' But even those who acknowledged, 'Yes, there are security risks,' there was a bit of paralysis about what to actually do about it."

As with so many other significant threats that were squeezed from view when the government went from panorama to a zoom on counterterrorism after 9/11, attention to climate-related impact on national security was demoted.

But Sikorsky says there also is a problem in the architecture of the national security machine, all across the executive branch.

"I think it's just really hard because of the structural way the intelligence community, in particular, was created," she said. "With such a regional focus, finding ways to slot in these transnational, functional issues has always been a challenge. And I think they've always played second fiddle, frankly, to the regional shops, and that's at the National Security Council, as well." There has not been an advocate focusing on climate change as a national security threat at the level there has been for Russia, China, or terrorism.

"I think you need to strengthen the seat at the table that the climate security folks have, to the level of weapons of mass destruction," she said. "Because the impacts are that great. Right?"

Intelligence analysts never get promoted for happy reports, but Sikorsky was unexpectedly upbeat about the government's ability to bring a cadre of next-generation leaders into the national security machine who have a clear-eyed understanding of the importance of climate-related threats, and it comes from her teaching responsibilities at George Mason University, in the Northern Virginia suburbs of DC.

"I see the newer generations of folks joining government and working in these agencies," she said. "I teach a course at GMU [George Mason University] on climate insecurity. Many of them are students in the military. I don't really need to explain to them why climate change is a security threat. They get it. They're like 'Yeah, of course it is. Because I've seen it on deployment when I was in Iraq.' So I think some of that culture piece is changing."

On October 7, 2021, the Defense Department offered its clearest statement ever acknowledging the national security risks of climate change to the country and to the military's ability to carry out its mission.

"Climate change is an existential threat to our nation's security, and the Department of Defense must act swiftly and boldly to take on this challenge and prepare for damage that cannot be avoided," said Defense Secretary Lloyd J. Austin III.[6] "Every day, our forces contend with the grave and growing consequences of climate change, from hurricanes and wildfires that inflict costly harm on U.S. installations and constrain our ability to train and operate, to dangerous heat, drought, and floods that can trigger crises and instability around the world."

Austin spoke in conjunction with the release of the Pentagon's new "Climate Adaptation Plan,"[7] which the defense secretary said would "be our guide for meeting the nation's warfighting needs under increasingly extreme environmental conditions—and for maintaining force readiness and resilience well into the future."

---

[6] "Statement by Secretary of Defense Lloyd J. Austin III on the Department of Defense Climate Adaptation Plan" (press release), US Department of Defense, October 7, 2021, https://www.defense.gov/News/Releases/Release/Article/2803761/statement-by-secretary-of-defense-lloyd-j-austin-iii-on-the-department-of-defen/.

[7] Department of Defense, Office of the Undersecretary of Defense (Acquisition and Sustainment), *Department of Defense Draft Climate Adaptation Plan* (report submitted to National Climate Task Force and Federal Chief Sustainability Officer) (Arlington, VA: Department of Defense, September 1, 2021), https://media.defense.gov/2021/Oct/07/2002869699/-1/-1/0/DEPARTMENT-OF-DEFENSE-CLIMATE-ADAPTATION-PLAN-2.PDF.

The very next day, we sat down for breakfast with Richard Kidd, the deputy assistant secretary of defense for environment and energy resilience, who described the scope of the problem and what the Pentagon simply has to do.

"No entity can opt out of the effects of climate change," he said.[8] "Climate change is going to be the context of the world that we live in from now on. Likewise, no entity can opt out of their responsibilities or requirements to take necessary steps at either adaptation or mitigation." Going forward, Kidd explained, "We're going to adjust and modify the existing programs in the department, whether it's military construction or some of our land management practices to adjust for climate change. We're going to adapt our training, we're going to adapt our plans, policies, and procedures. So we're going to pivot the entire department towards living and operating in a reality altered by climate change."

Officials concede that the effort is so all encompassing that many of the specific details remained to be worked out over coming months and years, but the action plan called on the Defense Department to focus on elevating its "climate-informed decision-making," to "train and equip a climate-ready force," and to build resilience into its installations.

Kidd said one significant focus of the department will be on reducing emissions. The American military burns more fossil fuel than any other institution on the planet, releasing more greenhouse gases than two-thirds of all other nations, the

---

[8] Richard Kidd, "Richard Kidd, the Deputy Assistant Secretary of Defense for Environment and Energy Resilience" (transcript), Defense Writers Group, Project for Media & National Security, George Washington School of Media and Public Affairs, October 8, 2021, https://nationalsecuritymedia.gwu.edu/project/richard-kidd-the-deputy-assistant-secretary-of-defense-for-environment-and-energy-resilience/.

Pentagon acknowledges.[9] The effort to combat emissions will require seeking alternative energy sources as well as greater efficiencies. That, in turn, requires cooperation with defense contractors, large and small.

At the same time that the Pentagon released its Climate Adaptation Plan, the Office of the Director of National Intelligence released the intelligence community's assessment of the risks, in a National Intelligence Estimate entitled "Climate Change and International Responses: Increasing Challenges to US National Security Through 2040."[10] And the Department of Homeland Security also released a report on climate-related risks[11]—marking the first collective statement on this threat from across the national security agencies.[12]

To be sure, there was more sense of purpose than specific, practical steps forward. And though the Obama administration had sounded the trumpet on risks from climate change, these statements were a coordinated and concrete sign of the

---

[9] Corey Dickstein, "Austin Calls Climate Change an 'Existential Threat' to US Security, Orders More Defense Department Planning," *Stars and Stripes*, October 8, 2021, https://www.stripes.com/theaters/us/2021-10-08/climate-change-defense-department-strategy-hurricanes-floods-fires-3173039.html.

[10] "Climate Change and International Responses Increasing Challenges to US National Security Through 2040" (National Intelligence Estimate, NIC-NIE-2021-10030-A, National Intelligence Council, Washington, DC, 2021), https://www.dni.gov/files/ODNI/documents/assessments/NIE_Climate_Change_and_National_Security.pdf.

[11] US Department of Homeland Security, *DHS Strategic Framework for Addressing Climate Change* (Washington, DC: US Department of Homeland Security, October 21, 2021), https://www.dhs.gov/sites/default/files/publications/dhs_strategic_framework_10.20.21_final_508.pdf.

[12] Christopher Flavelle, Julian E. Barnes, Eileen Sullivan, and Jennifer Steinhauer, "Climate Change Poses a Widening Threat to National Security," *New York Times*, October 21, 2021, https://www.nytimes.com/2021/10/21/climate/climate-change-national-security.html?searchResultPosition=2.

Biden administration's pledge to combat the impact of a more hostile global climate on national security.

The National Intelligence Estimate offered three key judgments:

1. "Geopolitical tensions are likely to grow as countries increasingly argue about how to accelerate the reductions in net greenhouse gas emissions that will be needed to meet the Paris Agreement goals."
2. "The increasing physical effects of climate change are likely to exacerbate cross-border geopolitical flashpoints as states take steps to secure their interests. The reduction in sea ice already is amplifying strategic competition in the Arctic over access to its natural resources. Elsewhere, as temperatures rise and more extreme effects manifest, there is a growing risk of conflict over water and migration."
3. "Scientific forecasts indicate that intensifying physical effects of climate change out to 2040 and beyond will be most acutely felt in developing countries, which we assess are also the least able to adapt to such changes. These physical effects will increase the potential for instability and possibly internal conflict in these countries, in some cases creating additional demands on US diplomatic, economic, humanitarian, and military resources."

For its part, the Department of Homeland Security—which includes both the Coast Guard and the Federal Emergency Management Agency—said its preparedness grants at the state and local levels will require an assessment of the impact of climate change, and it warned that expanding access to the

Arctic will certainly increase competition for fish and minerals. Nobody is yet calling it the new Great Game, but the major players operating in the Arctic—militarily and economically—are the United States, Russia, and China.

Interestingly, senior national security officials see in the risk of climate change a way for Western allies to blunt China's growing economic and military influence in the Western Pacific. China is all about expanding its influence and trade, and China is a huge polluter that has made little effort to incorporate mitigating climate change into its foreign and national security. In that gap is a space for the United States, Australia, and other allies to step ahead of China.

Just days after a new government in Australia was elected in 2022, the new deputy prime minister—dual-hatted as defense minister—visited Washington. Richard Marles previously held the government portfolio for Australia's relationship with the Pacific, so he knows the large nations and tiny island states well. He describes how families from the small island nations in the Western Pacific save and pool their money to send their children to school in Australia, knowing they may not have a home to return to as the ocean, little by little, swallows their land.

To compete with China in the region, it is up to Australia, the United States, and other partners to focus on "dealing with the issues that actually matter to these countries," Marles told us.[13] "First and foremost amongst them is the issue of climate change, which is felt around the world. But for the countries of

---

[13] Richard Marles, "The Hon. Richard Marles, Deputy Prime Minister & Defense Minister of Australia" (transcript), Defense Writers Group, Project for Media & National Security, George Washington School of Media and Public Affairs, July 14, 2022, https://nationalsecuritymedia.gwu.edu/project/the-hon-richard-marles-deputy-prime-minister-defense-minister-of-australia/.

the Pacific, for a low-lying atoll nation like Kiribati or Tuvalu or Marshall Islands, this is felt viscerally in a way which is hard to understand unless you've actually been there and seen what kind of, in an almost cultural sense, where the oceans have been a source of comfort, a source of food, a basis of culture, is now being seen as a source of anxiety and threat."

The policy focus for Australia, the United States, and like-minded nations, he said, should be "on our engagement with the Pacific on those terms—and in doing that, I am confident that we will be the natural partner of choice for the countries of the Pacific"—and not China.

Sharon Burke thought early and often about security and climate change, and her work earned her a position as the Obama administration's assistant secretary of defense for operational energy when that office was created. Its mission was to focus on the military's energy security.

Burke outlined for us some of the ways that climate change affects the US armed forces, starting with the operating environment.[14] "That would be the effect on installations—you know, they've twenty million acres under management worldwide. And some of those are coastal. Some are in arid places. Some are in permafrost, like in Alaska, locations where there are drastic changes. Some are in flood zones. Then California, of course, where there's been a lot of wildfires in recent years. All of that is a challenge for readiness."

The challenge is acute for the National Guard, the nation's citizen-soldiers. The National Guard mission today truly is biblical: fire, famine, pestilence. And add in a modern Horseman

---

[14] Sharon Burke (former assistant secretary of defense for operational energy), interview with authors, March 24, 2022.

of the Political Apocalypse, with urban unrest and riots. The Guard's orders for humanitarian support, natural disaster support, and support to civil authorities all are growing at the same time.

Gen. Daniel R. Hokanson, the chief of the National Guard Bureau, used to talk about the hot summer fire season. "Well, they kind of run year-round now," he told us.[15]

The Guard numbers 445,000 in all fifty states, three territories, and the District of Columbia, and one date pulled at random shows stress on the force. On Labor Day 2020, there were 64,000 members of the Guard on duty. About 20,000 were on missions overseas in thirty-four different countries. About 3,500 were on hurricane relief. About 1,500 were assigned to civil disturbances. Others on that day carried out a difficult rescue mission in Alaska and another team flew helicopters back and forth through a wildfire to rescue 240 people.

Each of those missions takes Guard members away not only from their families, jobs, and civilian lives—but also from deployment on other missions. That is only going to increase.

When Hurricane Ida formed in August 2021, it became the second-most damaging storm to hit Louisiana, after Katrina in 2005. But when Ida slammed the coast with Category 4 winds, Louisiana's largest National Guard unit, the 256th Infantry Brigade Combat Team, was deployed overseas. "Folks came from all the other states to help out," Hokanson

[15] Daniel R. Hokanson, "General Daniel R. Hokanson, Chief of the National Guard Bureau" (transcript), Defense Writers Group, Project for Media & National Security, George Washington School of Media and Public Affairs, November 10, 2021, https://nationalsecuritymedia.gwu.edu/project/general-daniel-r-hokanson-chief-of-the-national-guard-bureau/.

said. "And, in fact, Louisiana at that same time had one of their helicopters in California fighting forest fires." The relatively seamless movement of Guard forces into other states to backfill for local Guard members deployed elsewhere is a tribute to advance planning—but also a sign of increasing stress.

The National Guard is a hybrid force, funded both by the states and the federal government. The majority of its personnel hold civilian jobs and are on call by governors in case of emergency—but they also serve as a backup force for the president and the Pentagon when activated on federal missions. Assigning the Guard an increasing number of climate-related operations could cause strains and disruptions to the rest of the armed services and may prevent the Guard from fully serving its traditional role of "Strategic Reserve" for the Pentagon, in the view of experts like Burke.

Burke, currently a fellow in the Future of War Project at New America, a Washington policy institute, is concerned that the resources are not in place for the Guard to meet its historical missions and the accelerating operational tempo. "Is the Guard prepared to meet that demand signal?" she asked. "Do they have the people in the right place? Do they have the equipment? And then are they still going to be sufficient to have to be a war reserve as well?"

The Navy also faces a climate change threat beyond important real estate becoming submerged under rising seas. It's something straight from a science fiction novel, one set on a melting polar ice cap that releases horrific pathogens.

"In addition to the role of potential emerging diseases as the environment changes and perhaps new pathogens appear, we'll continue to reinforce our surveillance of emerging diseases," said Rear Adm. Bruce L. Gillingham, the Navy surgeon general.[16] "Our Navy medical research labs around the world are part of a global emerging infectious disease network that includes the World Health Organization and the Centers for Disease Control. So we'll continue to be very vigilant there."

Similar to the long buildup of the threat of a rising China, which had been warned for years, but without urgency, so, too, the way that climate change undermines national security has not been a classified secret. "I distinctly remember people saying, 'Yeah, that would be a problem in 2020 and 2030. It'll start becoming a problem,'" Burke recalled. But now it's 2022. "The frequency of these extreme weather events—it's happening."

Sherri Goodman is another person who was warning "way back when" about risks that are upon us today. In 2007, she wrote a landmark report, "National Security and the Threat of Climate Change," for CNA, a national security research center. The report is considered the first formal clarion call on the topic.

The military likes to talk about how it fields force multipliers to overwhelm an enemy with exquisite logistics and supply chains, advanced intelligence, surveillance and reconnaissance, and accuracy of weapons. Today, Goodman said, "you've got all of what I would call the climate threat multipliers. They all affect our base infrastructure in some way."[17]

Goodman, now a senior strategist and advisory board member at the Center for Climate and Security, compliments

[16] Bruce L. Gillingham, "Rear Admiral Bruce L. Gillingham, Surgeon General of the United States Navy" (transcript), Defense Writers Group, August 19, 2021, https://nationalsecuritymedia.gwu.edu/project/rear-admiral-bruce-l-gillingham-surgeon-general-of-the-united-states-navy/.

[17] Sherri Goodman (climate strategist and former Pentagon official), interview with authors, April 5, 2022.

the military for opening a campaign to make its bases and installations more climate resilient. And there is support in Congress, which is required, since change requires money. No member of Congress wants to see an important installation in his or her district closed or moved owing to the effects of climate. "So it's got wide bipartisan support," she said.

Goodman's career illustrates the slow and lumbering movement toward understanding the real threat. From 1993 to 2001, Goodman served as deputy undersecretary of defense for environmental security—"so I was the chief environment Safety Health Officer for DOD," she said. "And, you know, when I first got there in '93, the major issues were cleaning up contaminated bases. The first half of my tenure was really about coming into compliance with environmental laws and contaminated bases. And then working with Russia and former Soviet states cleaning up contamination from their industrial complex."

It was only in 1997, she said, in the run-up to the negotiations for the Kyoto Protocol, that the Pentagon began to focus on attacking greenhouse gases and how cutting those emissions would affect military operations. "We weren't really looking at any of the things like sea level rise and extreme heat," she said. "We were more looking at the mitigation side, like how would we control emissions for military activities? And, of course, there was natural resistance."

Goodman said the evidence has been clear, and for a long time. "You've got a planet that's like a patient who's been diagnosed with cancer," she said. "But you keep smoking, even though you know you've got cancer. You've got the diagnosis, but you're still smoking."

So, what is to be done?

Energy resources are a tool essential to military operations—but they also are a weapon, as we have seen with Russian threats to cut off natural gas supplies to Europe in retribution for sanctions over the Kremlin's invasion of Ukraine. Dependency is death for the military, so the Pentagon must accelerate efforts in the field of synthetic fuels, as well as solar power (many Marine units in the field light their tents this way), and longer-lasting, less-heavy batteries. Hybrid electric motors can offer a 20 percent savings in fossil fuel energy, Pentagon officials say, and so a shift is required.

These transformations will be no less dramatic than historic transitions from wind to coal and then coal to oil and then oil to nuclear.

Several military installations are experimenting with energy (and internet) independence, including major posts at Parris Island, South Carolina, and Miramar, California, and a smaller logistics hub in Albany, New York. They are creating independent, stand-alone microgrids and sufficient energy storage to allow them to unplug from the outside world and operate in a completely self-contained way—no outside power and no internet connections—for up to fourteen days. These systems would be a defense against an energy blackout—and in case of cyberattack—completely unplugged from the outside world's power grid and World Wide Web.

Lessons from these experiments could help the military create more independent operations in the field, lessening the logistics trains that offer tempting targets to an adversary as well as increasing energy consumption for transporting supplies over long distances.

Although the solution is obvious, it will cost billions of dollars to mitigate the effects of rising seas at coastal bases, wildfires in hot, dry regions that host military installations, and extreme rains and floods in the heartland.

No firm Pentagon-specific predictions are available, but the Office of Management and Budget released a government-wide assessment in April 2022 warning that, for the rest of this century, the US government could spend between $25 billion and $128 billion in extra funds each and every year to deal with six types of climate-related disasters: coastal disaster relief, flood insurance, crop insurance, healthcare insurance, wildland fire suppression, and flooding at federal facilities. In its report, the OMB offered these specific warnings of the trillions of dollars overall that could be required to deal with the effects of climate change and extreme weather:[18]

- Crop insurance premium subsidies are projected to increase 3.5 percent to 22 percent each year resulting from climate change–induced crop losses by the late century, the equivalent of between $330 million and $2.1 billion annually.
- Increased hurricane frequency could drive up spending on coastal disaster response between $22 billion and $94 billion annually by the end of the century.
- Rising wildland fire activity could increase federal wildland fire suppression expenditures by between $1.55 billion and $9.60 billion annually, the equivalent of an increase of between 78 percent and 480 percent, by the end of the century.

[18] Candace Vahlsing and Danny Yagan, "Quantifying Risks to the Federal Budget from Climate Change," White House, April 4, 2022, https://www.whitehouse.gov/omb/briefing-room/2022/04/04/quantifying-risks-to-the-federal-budget-from-climate-change/.

- Over 12,195 individual federal buildings and structures could be inundated under ten feet of sea level rise, with total combined replacement cost of over $43.7 billion.

Within the Pentagon, and across the warning and action machines, preparation for climate change challenges comes down to tyranny of limited time versus tyranny of limited dollars. Many still argue for investment in current risks, current operations, current wars rather than better preparing for future risks, future operations, future wars. There is more money going to climate-related programs today, but it is not yet a true sea change in shifting funds to radically reshape the military.

In assessing the flurry of national security and climate statements and initial actions from the Biden administration, Goodman gives the executive branch "a six or seven out of ten in terms of intent."

But, she said: "A plan without resources is hallucination."

PART IV

WHAT IS TO BE DONE?

CHAPTER 9

RETOOLING THE MACHINE FOR THE AGE OF DANGER

As World War II drew to a close, America's leaders knew they needed to create a system—a new set of machinery, if you will—for the world to come. The United States and its allies had prevailed, but it was clear that a retooling was needed to prepare the United States for a world that was, notably, to be defined by nuclear weapons. The many efforts that followed led to the creation of the modern warning and action machines. This involved the creation of the Department of Defense, the Central Intelligence Agency, and the National Security Council, among other changes. It has been updated and revised over the years but still looks very much like the system that was created in 1947.

To be sure, another set of important changes occurred after 9/11, leading to the establishment of the Department of Homeland Security, the Director of National Intelligence, and coordinating mechanisms like the National Counterterrorism Center. At the time of this book's publishing, there has been no large-scale terrorism incident in the United States since 9/11—incidents like the Boston Marathon bombing notwithstanding—and a number of potential threats have been disrupted.

That the warning and action machines have withstood the test of time says a great deal about the wisdom underpinning the original design. It is also a credit to various leaders who saw mounting evidence for needed change and were willing to act. Just in recent years, it became evident competition in outer space was growing and involved not just governments but also a host of commercial actors. The existing organization within the Defense Department was not keeping up, to say nothing of ensuring the United States had a chance to stay ahead. Thus, in 2019 the US Space Force was created, first drawing on units of the US Air Force much as the Air Force had drawn on the US Army Air Forces when it was created after World War II. Space Force now draws on talent and capability across the military services. Though still in its infancy, Space Force represents an important institutional reform, the kind that has been a US hallmark over the years.

Of course, we also know there are instances when the machinery has not kept up. Despite being the world's leader in communications technology, home to Hollywood and the biggest advertising agencies on earth, and creating Facebook, Twitter, Instagram, and the like, the United States government remains incredibly inept at communicating with the world and reaching audiences that see the United States as a model to emulate—despite our many internal divisions.

Still, in light of the new dangers that have emerged—China, Russia, germs, digits, drones, and storms, to name but several—it is time for a broader institutional review to see whether the larger warning and action machines are delivering in the ways that they should. The United States spends more than $1.25

trillion per year on security. We do not believe that sum is too much for a nation of our size that has important interests across the globe. We do, however, believe the American people should expect a proper return on that investment. By return, we mean that the proper pieces have been put in place to protect the American people, that the warning machine is tuned to the most important threats and that it adjusts and adapts as circumstances change, that the action machine is maintained in a ready state—a warm engine, if you will—and is tuned to both protecting lives and advancing the most important interests, and not just in the arena of traditional military affairs.

The dawn of the nuclear age brought the most comprehensive look at the warning and action machines. We are now witnessing the dawn of several new ages. We are only now beginning to understand what it is like to live when everything moves at the speed of light. Information technology has impacted our lives in profound ways, both for good and for bad. The ability to search for anything and find relevant information within seconds is something earlier generations could only have dreamed about. We also know that falsehoods in the information age can fly in a matter in seconds, but the time to repair them still is measured by the metrics of the industrial age.[1] The information age has changed how we interact and entertain, shop and eat, learn and love, grow and prosper. It has empowered governments and enfeebled them. Some thought it would be a source of empowerment—the great leveling that would unleash human potential. We also know it is being used as a source of control, not just in places like North

[1] This refers to Jonathan Swift's famous reflection that "falsehoods fly and the truth come limping after it; so that when men come to be deceived, it is too late; the jest is over, and the tale hath had its effect."

Korea but also across Russia, China, Iran, and other authoritarian states. It has been a place for theft on a scale never seen before, as General Keith Alexander reminds us. The internet also has been a force multiplier for stateless terrorist organizations, allowing them to plan attacks, spread propaganda, recruit new adherents, and raise funds from the relatively safe haven of cyberspace.

The biotechnology age has followed quickly on the information age. Unlocking the human genome in 2003 was just the beginning that led to remarkable discoveries in identifying and curing disease and allowed for such amazing feats as the development of a COVID-19 vaccine within a year of the onset of the first global pandemic in a century.[2] It is worth recalling that the search for a polio vaccine went on for decades while young and old were stricken and perished.[3]

But the same technology that can be used to save lives can be used to destroy life. When paired with modern information technology—especially information technology with a brain or artificial intelligence—the consequences could be horrific, as we have seen. The early architects of the nuclear era worried about a "doomsday machine." The architects of the information and biotechnology eras have the very same concern. We have yet to produce the modern versions of "game theory" and "deterrence theory" that were so important to the nuclear age and that we can use this time to ensure the survival of the

---

[2] "The Human Genome Project," National Human Genome Research Institute, https://www.genome.gov/human-genome-project.

[3] Anda Baicus, "History of Polio Vaccination," *World Journal of Virology* 1, no. 4 (2012): 108–114, https://www.ncbi.nlm.nih.gov/pmc/articles/PMC3782271/.

human species. There is an urgency for this type of thinking to happen and take hold.

Climate change poses its own risks, enormous ones. Each year that goes by brings more heat, more storms, more drought, and more fires. Humans, of course, are incredible adapters, but humans may not be able to adapt fast enough to the speed of change that is coming. As we have seen, the action machine is already at work making needed changes so it can operate in the face of new climate-related dangers. But when dealing with climate-related dangers becomes a full-time job—say, for local fire crews, parts of the National Guard, and the Federal Emergency Management Agency—who, then, is available to pitch in when these service providers are needed for their other jobs? We cannot expect the action machine to have multiple full-time jobs. That is a recipe for disaster.

Contending with new problems is one thing. Demonstrating that the warning and action machines can deal with existing problems is another. America's leaders like to say with pride that we have the best military in the world. They are right. But even the best military in the world can lose a war if it does not learn and adapt faster than the enemy we are fighting. The United States has a mixed record when it comes to learning and adapting. It did a good job of finding and killing terrorists after the 9/11 attacks, which required skill and adaptation and involved the use of new technology, especially drones. And it also involved new command procedures, developed in Iraq, to identify, target, and eliminate terrorist activity in very short time cycles.

US military forces proved much less capable in executing stabilization missions in Afghanistan and Iraq. Ultimately, it is nearly impossible to succeed when paired with weak local leadership, a lesson learned in Vietnam that had to be learned again in Iraq and Afghanistan. Supporting a partner is one thing. Taking on the fighting for that partner is quite another. The United States does not have a good track record in fighting other people's wars. It is a lesson we should learn and remember.

We hope the nation has learned another critically important lesson after Afghanistan and Iraq to be applied during this new age of danger: Militaries do not win wars. Allow us to repeat that: Militaries do not win wars. At best, militaries can vanquish rival militaries. Only governments, meaning the efforts of the entire government, can win wars by establishing a lasting peace or at least a long period of stability. The genius of the Marshall Plan for postwar Europe is the textbook example.

As the Defense Department was tasked with winning two Forever Wars, and the warning and action machines were preoccupied in the Middle East, we let our military advantage slip vis-à-vis China. The story is complicated, as we have seen, but the reality is the United States could lose a war over Taiwan if we had to fight it. Perhaps not this year or next, but over the coming years. There are a lot of people working hard to make sure that does not happen, and we have met some of them on these pages. But the institutions—the machine—does not show the sense of urgency needed to ensure that if the US military is called upon it will succeed. Or better yet, to ensure that none of our adversaries conclude that this is the week or month or year when they just might prevail against the US military. The prospect of failure is the single best form of deterrence. We need our potential adversaries to understand this idea. Nobody starts a war they think they will lose.

Andy Marshall was among the people who worried about China's emergence and what it means for global security. We spoke with him just a few months before his death in 2019. He was ninety-seven years old at the time. To Marshall, China was a preoccupation. He spent many of his waking hours pondering what the United States should be doing, how China would react, and how the United States should anticipate the ways in which China would react. Marshall thought about moves in multiples, much like a grand chess player. He was not alone, but there are not enough Andy Marshall's, and not enough of them thinking with a genuine sense of urgency.

Oddly, time might just be on our side when it comes to China. The prospect that China's population could collapse by half over the next forty-five to seventy-five years should be shocking to anyone thinking about China's long-term national potential. It must be chilling to China's leaders. A society that loses half its population in such a short amount of time will face a nearly endless series of internal challenges. China could collapse under the weight of its population implosion, to say nothing of the challenges China faces from its significant populations of ethnic minorities. And though the Biden administration's Inflation Reduction Act, passed in 2022, was only a down payment on slowing the effects of climate change, China has done far less. In addition to losing people, China will also lose territory as once inhabitable areas are vacated because of heat and drought.

Of course, there is a *but* to this story. China has embarked on the most ambitious military modernization effort since Nazi Germany's in the 1930s. Our purpose here is not to compare modern-day China with Nazi Germany, but it is to suggest China could upset a military balance that has existed in East Asia for several decades. The United States maintains enormous military wherewithal. Only a foolish adversary would conclude that it could confront the United States and defeat it with ease. But America's military advantage has slipped, and a war over Taiwan would be a close-run fight. Nobody we know with detailed knowledge of the situation thinks otherwise. There are promising ideas that could help restore America's advantage—ideas like cats and kittens and more—but the institutions that need to get those ideas from the whiteboard to the drawing board to the test range and into the field need to be moving at a speed that assumes time is *not* on our side. Not every answer must involve some form of technological breakthrough, what military analysts sometimes call "unobtanium," something akin to the search for the Holy Grail. Some of the answer is found in fixing known problems, applying known technology in new ways, and getting existing institutions to work beyond their traditional approaches. That is what Howie Chandler and Marty Neubauer concluded. The military services cannot let tradition get in the way of solving operational problems. The nation that put a man on the moon can meet this challenge, but not if we take decades to do so. The danger is too great.

We put the question of urgency to former defense secretary Robert Gates. He was clear and to the point. "The sad thing is, there are many different offices and entities in the Defense Department that have a chop [or a role] on any decision on procurement. You've got everybody from the general counsel to cost evaluation to the comptroller to the service chiefs.... And not one of them can make a final decision. Not one. The chairman [of the Joint Chiefs of Staff] has no money, so he can't do anything. So, what you end up with is that on any matter that requires urgency, it doesn't happen unless the secretary or the deputy secretary grabs it."[4]

---

[4] Robert Gates, interview with authors, August 23, 2022.

Gates went on to say that for matters with urgency it is all about people and accountability. The leadership needs to be clear about roles, and people have to be held accountable. Gates, of course, knows of what he speaks. When he realized the machine was too slow in taking action to protect troops from roadside bombs in Iraq, he took action to ensure a new armored troop-carrying vehicle was produced outside the usual system for research, development, test, evaluation, production, and procurement. He knew it would not happen, or would not happen in a timely fashion, unless he was directly involved. Robert Gates's formula is people and accountability.

On dealing with China's growing threat to Taiwan, the secretary and deputy secretary of defense need to make the machine work in the way it was designed to work. They need to identify the right people and hold them accountable. Find the people with the right ideas. Test the ideas. Discard the ones that don't work. Grab hold of the ones that do, and get them into the field. This sounds easy, and we know it is not. But it is what needs to be done.

Slowness was a recurring theme in our discussions with current and former government officials. It worried General Hyten and Lieutenant General Hinote, and many others. The Cold War brought a sense of urgency to government decision-making. There was a single competitor, a pacing threat. When it comes to military competition, China is seen as the pacing threat. But the sense of urgency that existed in the Cold War, the idea that the American genius would be unleashed on solving the hardest problems, has not yet taken hold. The problems we confront today need minds like Tom Hamilton's working on the hardest problems. Not just tens or hundreds of Tom Hamilton's but thousands of them. The talent is out there. It needs to be harnessed. It needs to happen at a pace that keeps us ahead of the problems, not chasing them. The talent will not all be in the government or even most of it. It will be in industry, academia, think tanks, and the like. It will be across partner nations. It might even be in some person's garage. The key is to find the talent and harness it.

None of this is to suggest the problems are insurmountable. Russia's war in Ukraine has shown the vulnerability of modern military forces, even forces thought to be proficient and with relatively recent combat experience. War carries with it enormous uncertainty. Few thought Ukraine's military could withstand an onslaught from Russia, though some close military observers doubted that Russia could ever succeed in invading and occupying Ukraine with the size of its initial invasion force. We spoke with more than a few analysts who noted that even if Russia had succeeded in its initial attack against Ukraine's capital Kyiv, it did not have sufficient forces to occupy the country. It is a lesson the United States learned in the immediate aftermath of its 2003 invasion to topple Saddam Hussein in Iraq. The Bush administration had to watch its initial triumph collapse into chaos because the invading force was not capable of bringing stability to Iraq. As of this writing, Russia's military is now bogged down in Ukraine, and more than a few analysts have concluded it may ultimately lose the war. Whatever the coming months bring, Russia will certainly confront what former Secretary of Defense Donald Rumsfeld called "a long hard slog." No doubt that members of Putin's inner circle are conjuring similar thoughts. They should be reading Rumsfeld's October 2003 memo to his top advisers.[5]

---

[5] Donald Rumsfeld, memo to Gen. Dick Myers, Paul Wolfowitz, Gen. Pete Pace, and Doug Feith, Subject: Global War on Terrorism, October 16, 2003, https://irp.fas.org/news/2003/10/rumsfeld101603.pdf.

The war in Ukraine invokes another lesson. Despite an initial slow response from the Biden administration, which was understandable given that Russia was brandishing the prospects of nuclear weapons' use, the warning and action machines have shown remarkable sophistication in supporting an under-resourced but highly motivated Ukrainian partner in confronting a more powerful Russian foe. Moreover, the warning and action machines swung into action to build and sustain allied support for Ukraine. NATO's response, and Germany's, in particular, the so-called Zeitenwende or epochal moment, went well beyond what some observers had expected.[6] Indeed, Germany elected to shut down the Nord Stream 2 natural gas pipeline, which would have doubled the amount of natural gas being delivered from Russia to Germany. Russia turned off Nord Stream 1, and Germany is holding fast. Germany took an additional step, that few expected, committing to increasing its defense expenditures over the coming years. Putin's direct threat to Ukraine, and potentially Ukraine's neighbors to the west, persuaded Germany to take action that a succession of US presidents could not.

What is more, Russia's invasion prompted Sweden and Finland, which long had close relationships with the West but which sat outside the NATO alliance, to seek membership in the alliance. Membership was offered to both nations in July 2022, which brings significant additional political and military strength to the alliance and certainly changes Russia's geostrategic outlook.

Gates made this very argument in our conversation with him. Regarding Putin's calculations in choosing to invade Ukraine: "He wanted to weaken NATO, he's ended up tremendously

---

[6] Rachel Tausendfreund, "Zeitenwende—the Dawn of the Deterrence Era in Germany," German Marshall Fund, February 28, 2022, https://www.gmfus.org/news/zeitenwende-dawn-deterrence-era-germany.

strengthening NATO, and not just in resources and determination but having the Germans take the initiative and turning off the Nord Stream 2 pipeline.... Not to mention Sweden and Finland because they bring serious military assets to the alliance."[7]

Determined NATO. Expanded alliance. Resilient victim. Not the outcome Putin was seeking at all.

Of course, the war in Ukraine is far from over, and Putin continues to threaten the use of tactical nuclear weapons. But it is not too early to declare that Putin miscalculated, and the NATO alliance showed it still has the mettle that was such a source of pride through the Cold War. What we see in this illustration is that the warning and action machines still have the wherewithal to swing into motion when called upon. All was not perfect in the response, but more things seemed to go right than wrong, and the warning machine, in particular, was used in novel ways that helped provide a coherent response. Perhaps not since the time when Adlai Stevenson was dispatched to the United Nations with photos of Soviet missile construction in Cuba had intelligence been used with such dramatic effect to galvanize and sustain an international response.[8] For example, the CIA's deft use of previously classified intelligence to accurately warn Ukraine, and the world, of the pending Russian

---

[7] Robert Gates, interview with authors, August 23, 2022.

[8] James M. Lindsay, "TWE Remembers: Adlai Stevenson Dresses Down the Soviet Ambassador to the UN (Cuban Missile Crisis, Day Ten)," Water's Edge (blog), Council on Foreign Relations, October 25, 2012, https://www.cfr.org/blog/twe-remembers-adlai-stevenson-dresses-down-soviet-ambassador-un-cuban-missile-crisis-day-ten.

invasion went a long way toward rallying global support for Kyiv.[9] The machine still works when we need it.

When it comes to new dangers, it is important to realize it is not just the military that needs to be ready. The early days of the COVID-19 response revealed what happens when you are playing with a pickup team. The US military learned in Korea it has to be ready when the call comes in. Not tomorrow. Not next week. Ready today. That means no pickup games. This requires a steady diet of money and training—money to maintain a standing force and training to be sure it is ready at a moment's notice. The Army and Air Force have a saying that has permeated their institutions: "Ready to fight tonight." It is their way of saying they need to be ready the moment the call arrives. Tonight.

What is common across all the new dangers we have explored—germs, digits, drones, and storms—is that there needs to be a warning machine that is on par with the best warning the nation receives on military threats. Some threats move quickly—the major storm or fire that is ravaging an area, or the cyberattack that is crippling the East Coast gasoline supply—and others move more slowly. A world-class warning machine needs to be able to account for all types of

---

[9] It also did much to scrub away the lingering stain of the UN briefing Colin Powell, serving as George W. Bush's secretary of state, gave in making the case for invading Iraq. The intelligence said Saddam Hussein had an active program for weapons of mass destruction and that he refused to open his country to inspection. The threat, according to the intelligence, was clear and urgent: Invasion was required. As is painfully well understood today, the intelligence underlying that briefing was false. (Powell later confided to interviewers that he demanded that George Tenet, then the CIA director, sit directly behind him during the UN briefing, televised live around the world, to foot-stomp that the secretary of state was describing intelligence gathered by the CIA.)

threats—fast and slow, close and far—and provide decision makers with the type of information that allows them to put the action machine into motion. Generic warnings are one thing. Actionable intelligence—warning that can lead directly to action—is another. That means warning needs to come in a coherent fashion. Not dribbling in from hither and yon, but pulled from all available sources, carefully assessed, and packaged with intent.

The intelligence community once had a warning officer whose job it was to sound the alarm when a threat appeared to be imminent. That job was eliminated nearly a decade ago. We met with the last warning officer, the last Jedi, if you will. His name is Ken Knight, and he was a career intelligence officer. What impressed us about Knight and his job was that he was the one person who could break the glass in an emergency and sound the alarm. He could issue a warning of a pending problem, and with certainty it would land on the president's desk. It would also land on the desk of everyone else who reads what the president reads. When Knight was in his job, he would produce a half dozen or so warning notices a year. He did not always get it right. And he is humble enough to recount his misses, like the 2011 fall of Egypt's Hosni Mubarak. He recalled that the signals were all there, even the taxi drivers were complaining, but the intelligence community thought Egypt's military would rally around Mubarak.[10]

There are pros and cons surrounding the idea of a warning officer. Some argue that warning should be everyone's responsibility, and housing it in a single office or function takes the

[10] Ken Knight (former national intelligence officer for warning), interview with authors, March 2018.

specialist community off the hook. Others point out that specialists can become too enamored with their subjects and may not see the big changes or threats when they arrive. Knight did not push the idea of a single node or multiple nodes for warning. What is important is that warning be treated as a serious full-time job. There is an entire discipline needed to provide decision makers with adequate warning that can be acted upon. Warning is taught as a discipline at the National Intelligence University—yes, there is one. When we look at new threats like germs, digits, drones, and storms, we see the need for a warning community to coalesce and form new disciplines in much the same way that has been done with more traditional national security threats. We also need to recognize that warning of these threats will not necessarily come from the traditional intelligence community. Some of the warning will be sitting there in plain sight—the kind of open-source intelligence that sits out there for all to see, though not all will understand the importance of what they are seeing.

Ken Knight had another important idea to share. He talked with us about the distinctions between secrets and mysteries. In his use of the terms, secrets are matters that are hidden but that can be found. The spy trade relies on tools to uncover them. Humans and technology—and a blending of the two, often called sources and methods—comprise most of the tools. Watching, searching, listening, and comprehending, these are all responsibilities of the warning machine. They depend on data and analysis. They are bound by rules of evidence.

Mysteries, on the other hand, are matters that are unknown and may be unknowable. The warning machine deals with mysteries by developing hypotheses—often as if/then propositions. If this appears, then we should expect that to happen as a result. Knight cautioned that the warning officer cannot be held to the same rules of evidence when it comes to mysteries. Although we spoke with Knight long before Russia invaded Ukraine, he made a point of saying that even Putin may not know his next move in a crisis. Dealing with mysteries, or perhaps riddles, requires deep knowledge and constant attention. It requires the unblinking eye that is focused continuously on the problem at hand. Dropping in and out on occasion means you are likely to miss important signals and subtleties that will only be understood by the trained observer.

As we organize the warning machine to deal with new problems, it will need to be arrayed in a way to deal with both the secrets and the mysteries that surround them. It needs to identify the standards of evidence that will prompt decision makers to put the action machine into motion. It needs to develop and test the hypotheses that bring more clarity to the mysteries that surround these new threats.

New mechanisms may also be required. Just as the National Counterterrorism Center (NCTC) was created after 9/11 to ensure there was not a person in the White House with a phone on two ears—one from the CIA, another from the FBI—it may be necessary to have similar standing activities for germs, digits, drones, and even storms. The problem with these threats, as we have seen, is that authority runs not just across the federal government but also down to states and localities and even to the private sector. Instead of a phone on two ears, it is easy to imagine needing five, ten, or even fifty or more phones, given the very separate nature in which states act. Coordinating centers like the NCTC do not solve the "who's in charge?" problem we uncovered, but they can serve as a place to pull information together and get it into the right hands. There is no getting around the federal nature of the US system, but common standards and protocols can help ensure all are speaking the same language and acting with the best available information. National centers for biosecurity (including pandemics), cyber, and climate would be a step in the right direction.

Stephen Hadley was deputy national security director when the NCTC was first stood up. He later became the national security adviser in the second term of the George W. Bush administration. We talked to him about how the NCTC operated, and he reminded us that the original idea for the NCTC was not just to integrate intelligence but also to coordinate operations. It was to be the place where intelligence, military, and financial tools could be put together to fight terrorist threats, both people and organizations. He thinks it would be wise to not give up on the idea of fusing intelligence and operations as we consider NCTC-like functions for new dangers.[11]

In a much larger sense, when it comes to new dangers, the warning and action machines need to move beyond the boom-and-bust cycles that are so disruptive to the larger system. More funding is not the answer to every problem. But big increases and decreases in funding are a problem. Workforces cannot be built and sustained. Technology and equipment cannot be kept ready. Supplies cannot be kept in good order, or even kept. Expertise cannot be cultivated. And the machine itself is unlikely to be tested to determine whether it is up to the task. Just as the military runs drills and exercises to check on its readiness, so, too, should all parts of the larger warning and action machinery. Pickup games are something to be avoided.

---

[11] Steven Hadley (former national security adviser), interview with authors, September 16, 2022.

Leon Panetta offered the concept not of creating another new, large bureaucracy to manage emerging threats but to have assets within all the departments and agencies identified and on standby to move into action when needed to deal with specific risks.

Panetta said it is worth considering "some kind of task force that combines intelligence, law enforcement, and all of our domestic operations that focus on emergency protection," he said. "A task force that meets regularly, that includes all of these key elements, I think would be more effective in terms of protecting the country."[12]

Panetta may be best known for having served as CIA director when Osama bin Laden was found and killed, before moving over to be defense secretary. But he also has experience as chairman of the House Budget Committee, White House chief of staff, and director of the Office of Management and Budget. He is a classic Washington deal maker, fixer, and enforcer.

He noted that ensuring proper funding is not just a role for the executive branch. Congress has the power of the purse, and the relevant oversight committees have a responsibility to ensure proper levels of funding are being authorized and appropriated to contend with new dangers. Congress has played critical roles in reforming the warning and action machines over time, including the most recent reforms in the aftermath of the 9/11 attacks. Congress must play an equally important role in retooling the warning and action machines for this era. So far, Congress is barely keeping up. Others would say it has fallen way behind.

---

[12] Leon Panetta (former CIA director and former secretary of defense), interview with authors, June 2019.

Why? It is easier to beg Congress for money to fight traditional wars than to battle less-specific, less-visible, slower-moving threats. Just under three thousand Americans died on 9/11, and the nation launched two wars and spent trillions of dollars in a mission that began as counterterrorism and ended in failed efforts at democratizing Afghanistan and Iraq. More than a million Americans have died from COVID and, as we have written, the response began from a standing start and has stumbled along the way.

Threats should not become spending priorities only when they become military priorities. The nation's policymakers and its citizens need to expand the definition of national security far wider than just those risks that threaten to blow us up. We must have a better understanding, too, that future threats are not the kind that can be defeated, only managed, and those efforts will require continuing commitments of funds to such unheralded projects as the National Bio and Agro-Defense Facility at Kansas State University. Managing threats does not have the same panache as fighting and winning wars, but it is how the nation will need to deal with the new dangers.

Steve Hadley had a slightly different take, but one very much in concert with this idea. He mentioned a paper he wrote on reforming the National Security Council system, noting in particular that the current National Security Council structure, with its many committees and subcommittees, simply cannot cope with the vast array of national security, homeland security, and foreign, defense, and economic policy challenges that the United States faces in the world today. He mentioned the current system was built by Henry Kissinger at a time when problems were generally slow and predictable—not always slow and not always predictable, but generally so. Today, almost nothing is slow, and few things are predictable.

Hadley argued that the bandwidth of that system is too limited to cope with today's wide range of challenges. Under the current system, by the time an issue gets to the president's desk, it is often at the point of crisis. That leaves the president with the choice of breaking the glass and pulling the alarm—often by using the military—or doing nothing. The military likes to talk about having options "left of boom"—that is, to have options before something blows up. The president needs more options left of boom, especially when the dangers have such enormous consequences.[13] Hadley goes on to offer a host of practical recommendations. His overall message is that retooling the warning and action machines needs to begin at the top.

Hadley had an interesting point to make about people, too. He leans toward having fewer people at the top rather than more. Of course, we cannot have every problem managed by a single person or even a small handful of people. But Hadley's point was different. He allowed that the more specialized a person becomes, the less likely that person can see connections across problems or perhaps see opportunities inside problems. He wants a smaller number of people who can see connections and be ready to act upon them.

In this regard, more than a few of the experts we met with suggested a Goldwater–Nichols–like review of how the executive branch organizes its interagency activities. Just as Goldwater–Nichols clarified roles and responsibilities within the Defense Department, there is similar need to look at roles

---

[13] Nick Burns and Jonathon Price, *America's National Security Architecture: Rebuilding the Foundation*, Aspen Strategy Group Policy Book Series (Washington, DC: Aspen Institute, 2016), 107.

and responsibilities across the executive branch. Each of the dangers we examined cuts across much of the executive branch of government, and several involve states and localities. It is clearer to see who is in charge of Russia or China than it is to know who is responsible for threats to biosecurity and public health, cyber, climate, and drones. Even China presents enormous challenges. Hadley reminded us that China cuts across every functional area of government, from national security to the economy and just about everything in between. A coherent China policy will require enormous coordination, perhaps orchestration, across the whole of government and the private sector. At times, it will involve states and municipalities, to say nothing of coordination with allies and partners.

It is often said that many hands make light work, but too many hands can simply produce chaos. We have enough chaos. What we need instead is coherence—from warning through action. We need a lifesaving national security machine that complements our industrial-strength killing machine.

Somewhat paradoxically, it was this idea—the notion of a lifesaving machine—that led us to General Stanley McChrystal. It was McChrystal who became the face of the Iraq War when he provided regular briefings at the Pentagon during the early days of the invasion. He then became a legend of the Iraq War when he overhauled the military's Joint Special Operations Command (JSOC), in charge of the military's most elite counterterrorism forces, including SEAL Team 6 and Delta Force. He transformed the community of American commandos from a temperamental sports car rarely taken out on the highway into an industrial-strength killing machine never seen before in military history.

The JSOC that McChrystal inherited was designed to rescue hostages or, as he relayed to us, to go after terrorism's

Mr. Big. What he realized when he took command was that his bullets didn't have a target, and even if they did, it took too long to reload. Mr. Big always seemed to be beyond his reach. And some terrorist networks didn't even have a Mr. Big but a series of midlevel enablers, call them Mr. Littles, the sum of whose efforts in recruiting and planning and fundraising proved more tenacious than those of a single Mr. Big.

McChrystal's challenge was how to fix the machine that was used for hunting terrorist targets. It was too slow and stodgy to be successful against a thinking, adaptive network. He needed to create a new system. He needed to get inside the operations, inside the patterns of life, inside the decision cycle of the terrorist organizations. He had to find ways to harness technology, to bring together all the available information and use it in novel ways. More importantly, he needed to put his own troops into action to create new information, which he called the fight for intelligence. He built a system where individuals up and down the chain of command could make decisions so that many decisions could be made in a day. He established rules but allowed his subordinates to act. "We built it. We managed it. We watched it. We didn't screw with it. We weren't in the machine," he said. "We built processes and we set up rules. Nothing illegal. Nothing immoral. Then we let it run." He later came to have an expression for describing his system: "Eyes on, hands off."

Ultimately, he needed to create a system that allowed him to poke and prod inside his enemy's networks, to create waves and reverberations that would, in turn, generate new information on the next series of targets. Ways to dislodge the secrets Ken Knight talked to us about.

And in what was his most important decision, he broke down traditional bureaucratic walls that for years had allowed the CIA to carry out national security missions its way, without sharing. The FBI did the same, without sharing. And both rarely had conversations with Treasury or other agencies with lines on how terrorists finance their violence. He brought representatives from each into his headquarters and bound their efforts to support his fighters, making all of them loyal partners.

The JSOC that McChrystal inherited was capable of about four operations a month (the kind we read and hear about and see depicted on television and in the movies). The JSOC that McChrystal and his team produced—a system built for stimulus and response, warning and action—was capable of ten operations a night, or three hundred a month. It was focused on fusing warning with action, which in turn yielded more warning and more action. Often, intelligence gathered from one raid—documents and computer drives—would be immediately scoured for clues and used to direct a follow-up raid even before word of the first raid had reached terrorist planners. McChrystal called it "the fight for intelligence." He put it to work and in time it yielded Mr. Big—it led to the death of top terrorist leader Abu Musab al-Zarqawi in Iraq and eventually found and finished Osama bin Laden in Pakistan. His JSOC rewrote the military manual and made history.

In early May 2020, we talked with McChrystal about the crisis engulfing the country—the virus that would kill over a million Americans. The death toll from COVID has surpassed American losses in World War II, Korea, Vietnam, Afghanistan, and Iraq. Americans became accustomed to hearing death toll tallies on the evening news. Microbes had proven far, far deadlier and more indiscriminate than terrorists.

We wanted to know how the guy who built an industrial-strength terrorist killing machine would solve the crisis. He suggested the nation build an industrial-strength lifesaving machine.

McChrystal is even-tempered and unassuming. Even though we both knew him from his time in the military, he insisted we no longer address him as General but call him Stan. Our conversation began early in the morning. Military leaders almost always start early. In fact, there's an expression often touted in military circles, "If you're not early, then you're late." Because we were still honoring social distancing protocol, we met by phone rather than in person. McChrystal was true to form. He dialed in three minutes before we were scheduled to begin.

We talked about the role of the federal government as a synchronizer and coordinator. We talked about the idea of having a national policy but pushing authority down to the lowest level where it made sense to make decisions, of how to connect national, state, and local decision makers and mechanisms. McChrystal believes in "eyes on, hands off" and keeping routine decisions out of politics. Warning and action need to be paired. When certain types of warnings surface, very specific actions should follow. Alerts should go out. Stockpiles should be checked and topped off. Orders should go out. Equipment checked and moved. People alerted. Reserves called up. If the warnings intensify, additional actions should follow. Trained specialists should be mobilized and deployed. Local facilities should be put on standby for emergency conditions. Contracts should be let. Resupply should be ordered. Supply chains put into motion. Personnel moved from one location to another.

McChrystal compared the system he was imagining to the current defense readiness condition, or DEFCON, that guided military preparations during the Cold War. In McChrystal's vision, top leadership—the secretary of defense or president—determines the DEFCON. Once established, a whole set of preplanned actions fall into place. Said differently, once the political leadership determines the severity of the condition and the threat, the system is designed to act in very specific ways. The DEFCON changes, increasing or decreasing, and the system acts accordingly. But there are plans in place so that within each level of readiness, individuals are empowered to take specific actions. They don't need approval or permission to act. They are required to operate within bounds, but they are empowered to act. The most sensitive actions and weapons are protected by fail-safes. "Eyes on, hands off."

There are risk conditions, as we have seen, for an entire array of possible threats: To health. To the food chain. To the energy grid. To the water supply. To communications and cyber networks. To our elections. Think of establishing national standing joint task forces that link all parts of the national government and that include states and municipalities as partners in establishing a series of flexible plans in case the risk levels rise in HEALTHCON, FOODCON, ENERGYCON, WATERCON, CYBERCON, ELECTIONSCON. It would require a significant government restructuring, but no more than those accomplished in 1947 and post-9/11, to create an agile system with the talent, expertise, and authority to prepare for the entire range of risks, and not just a terrorist attack on the homeland.

The McChrystal concept would build several JSOCs for protecting the American public from the next pandemic, the next crippling cyberattack, the next crop blight, the next systemwide crisis that will surely come along and that cannot even be predicted today. The idea is similar to Panetta's suggestion of identifying elements across the security bureaucracy to serve on standing joint task forces to confront with agility rapidly emerging and unpredictable threats.

One of the reasons Steve Hadley likes experiments with managing policy areas is that, for some of the issues, the connections have never been made, the people and organizations have never been assembled. He pointed out, "For some of these issues, you're talking about you're going to need the domestic issue agencies, and that gets you to your whole of government. But for some of these things, particularly cyber, you need whole of society, because the key actors are in the private sector, and that kind of joint task force approach to involve not only all of government but also the private actors has never been done. We don't know how to do a whole-society exercise."[14]

The time to experiment is before the crisis arrives. We are almost certain to not get it right the first time around.

Hadley also had some cautions. We talked about possible risks of involving the private sector in the shaping and making of policy in a crisis. One of the reasons Operation Warp Speed produced a COVID vaccine in such little time is because the secretary of health and human services invoked the PREP Act—public readiness and emergency preparedness—that provided liability protections to manufacturers, distributors, states, localities, and licensed healthcare professionals, among others. It was invoked in March 2020, just as COVID was leading to the shutdown of the American economy. It did not absolve anyone of willful misconduct, but it did allow Warp Speed to avoid the kind of delays associated with typical bureaucratic red tape.[15]

---

[14] Steve Hadley, interview with authors, September 16, 2022.

[15] "PREP Act Immunity from Liability for COVID-19 Vaccinators," US Department of Health and Human Services, last reviewed April 13, 2021, https://www.phe.gov/emergency/events/COVID19/COVIDVaccinators/Pages/PREP-Act-Immunity-from-Liability-for-COVID-19-Vaccinators.aspx.

Hadley applied this same logic to cyber. He noted, "If the private sector is going to share their problems with the government national security agencies to get help in fixing them, you can't give it to the SEC and let them call them out for failure to disclose it in their annual reports." He went on to make a broader point: "So, there are entry prices that have to be paid. If you're going to get the private sector to the table, liability from third parties is one. Liability from government action and government enforcement action in other sectors is going to be another."

Hadley's key takeaway was this: "It's what you've got to do. You've got a national emergency. You've got to take action like that." It sounds a little like the military's version of ready to fight tonight.

We asked Hadley where he would house the experiments. He reflected for a moment and conceded they probably need to be housed in the White House. Not for all time, but for long enough to get them off the ground. He talked about how the President's Emergency Plan for AIDS Relief (PEPFAR) was given life in the White House and was transferred successfully to the State Department. Though not as well known domestically as other decisions of the Bush (43) administration, the PEPFAR program is widely praised across Africa, even today. But Hadley acknowledged that Condoleezza Rice was national security adviser and went on to become secretary of state. She wanted the program to succeed at the State Department, and so it did. Other initiatives handed from the White House to departments and agencies did not succeed nearly as well.

It is with this in mind that the idea could go a step further. One senior official retired from White House inner circles suggested having a pair of presidential national security advisers, one to deal with the near-term threats and one assigned to look around the corner and over the horizon. In military speak, you always fight the crocodile closest to the canoe. So, who is looking downriver to warn of that gigantic waterfall? And is that person empowered to walk into the Oval Office and focus the president's attention on a future threat when the president's inbox already is overflowing? Too many cooks? Who knows. But perhaps another experiment worth considering.

Here is what we do know. One official who served in senior national security positions in multiple Democratic administrations offered a caveat about presidential decision-making that is as practical and honest as it is concerning. Although a president's national security team will describe policy options on the basis of their best assessment of opportunity, risk, and outcomes, a president's political team, whose focus is on domestic politics, especially electoral politics—and likely voter response to any national security decision—always has the president's ear as well. The tension between national security needs and electoral needs is certain to increase as the nation's politics become even more polarized.

One issue that remains completely predictable is the need for the United States to nurture and expand its alliances to forge a strong and unified response to this new age of danger. The Biden administration deserves credit for rebuilding NATO after the damage of the Trump years. It also has expanded cooperative deals, such as the British–American agreement to sell Australia fast-attack, nuclear-powered submarines to deter an expansive China, what is known as the AUKUS agreement. An important but less-noticed aspect that underpins the AUKUS agreement is technology sharing. America's safety will require building more international partnerships with technologically advanced democracies like our own to prepare for the known threats and the still unknown. There are many willing and able participants, but the United States will have to lead. It will take time and require persistence. Some efforts won't succeed, as we learned in our earlier efforts at building alliances. But common interests will form the glue of the new partnerships that are needed. We see it happening in Europe, with Finland and Sweden seeking NATO membership. We see it happening in Asia, with the United States, Japan, India, and Australia engaged in what has become known as the Quad, the brainchild of Japan's former prime minister the late Shinzo Abe to create an "Asian arc of democracy." It is still very much an informal mechanism that is taking on a host of important issues.

Perhaps a coalescing idea is one offered by Paul Stares, director of the Center for Preventive Action at the Council on Foreign Relations, an American think tank. By reassessing the Cold War nuclear suicide pact with the Soviet Union called mutual assured destruction, or MAD, Stares suggests that the United States lead an effort to create an international system of "mutually assured survival" so that the United States and its allies can deal with the threats of China and Russia.

"It has broader meaning than purely just demonstrating or reassuring each other of benign intent," Stares said. "Unless we reassure each other that we are not seeking to, essentially, completely undermine the political basis of the ruling Communist Party in Beijing and Putin's role in Moscow, then you're not going to get any productive restraint on these other military developments, as well as on common security concerns."

Further reframing the Cold War vernacular, Stares offers the idea of "collective coexistence," different from the Cold War idea of "peaceful coexistence," which presumed that we will be mutually respectful and appreciative. Instead, future relations between the West and Beijing and Moscow could be based on shared security predicaments—climate change, terrorism, crop blight—that are threats to all without respect to borders. "And that's not to say that we are completely mute about our criticism, that we don't hold back," Stares said. "It comes down to reassuring each other about political intentions towards one another." Note to Putin: No more election hacking if you want respect from the United States.

Survival will require concerted action not just within the government and across the private sector but also with allies and partners and potential adversaries. Steve Hadley talked about a whole-of-society approach. It will require a whole-of-world approach, too.

The future needs a seat at the table. Time and again, our nation has let the future catch up with us. Retooling the warning and action machines to meet this new age of danger is a matter of urgency. How to do it? If recent decades have taught us anything, it's that the seemingly urgent has a way of displacing the quietly important. The immediate overshadows the pending. Not always, but often enough. The future always comes calling, and attention must be paid to creating the warning and action machines we need for this era of new superpowers, new weapons—and new threats, many of which will be defined not in traditional military terms. We should not have to wait for the next war or pandemic or massive cyberattack or climate-related calamity to have the needed machinery in place.

Lest we forget, the future always leaves a calling card.

ACKNOWLEDGMENTS

This book germinated about five years ago over a long meal at a restaurant near the Pentagon, when we discussed national security threats crouching just over the horizon and receiving insufficient attention—ticking time bombs, if you will. Over many months, and then years, that idea focused and, we hope, sharpened into the book you are reading now.

In many ways, however, this book first began taking shape more than thirty years ago. Andy was a Pentagon strategist helping shape post–Cold War defense strategy. Thom was a reporter in Moscow, watching and writing as an empire crumbled and a new order emerged. Andy and Thom got to know each other a little over twenty years ago when Andy was head of strategy at the Pentagon and Thom was covering the Pentagon for the *New York Times*. We have been close colleagues since.

There are many people who deserve our sincerest thanks.

Sonni Efron was present at the creation and helped shape our early thinking.

Sam Raim, our first editor at Hachette, enthusiastically understood what we wanted to say and brought his own line of questions that drove us to elevate our thinking. Dan Ambrosio carried our manuscript through final edits with care and professionalism and offered important advice on polishing the final draft. Thanks also to Alison Dalafave for her support, and to Sean Moreau, the production editor for this book, and Christina Palaia, our copyeditor, for caring about our text with smart questions and powerful attention to detail.

We offer a special thanks to Alex Ward, the former director of book development at the *New York Times*, who read our proposal and gave us a swift kick in the southern hemisphere just when we needed it.

And, of course, we offer heartfelt thanks to our literary agent, Bonnie Nadell, who shepherded this project from inception to publication.

Our manuscript received thorough and thoughtful critiques from Kori Schake, director of foreign and defense policy at the American Enterprise Institute, and from Michael O'Hanlon, director of research in foreign policy at the Brookings Institution. This book also benefited from interviews, conversations, and critiques from countless colleagues across the wider US national security community, only some of whom appear on these pages. We owe a debt to all who provided us support along the way. Any remaining flaws are solely our own.

Andy would like to thank Michael Rich, RAND's former president, and Jason Matheny, RAND's current president, for their steadfast support. Andy owes a special debt to his many RAND colleagues who were quick to answer questions or help him ponder riddles. It is hard to imagine a better collection of talent anywhere. Tristan Finazzo provided valuable research assistance. Jeff Hiday provided valuable comments as our writing process was drawing to a close.

Thom would like to thank Silvio Waisbord, Frank Sesno, Sarah Morrisette, Maria L. Jackson, and Susan Boerstling for their unwavering support of Thom's work directing the Project for Media and National Security at George Washington University's School of Media and Public Affairs. Several on-the-record discussions with senior national security officials hosted by the project contributed to the content of the book, so Thom would like to thank Carnegie Corporation of New York for its generous philanthropic support of the Defense Writers Group and related programs. Thom also thanks the Howard Baker Forum for its generous support of the Cyber Media Forum.

As we complete this book in early 2023, we are acutely aware that trying to capture in place a world undergoing constant and significant change is like trying to paint a bullet train. Whatever the shape of the national security landscape upon publication in the spring of 2023, we hope the larger questions we address, and attempt to answer, remain meaningful amid global changes large and small.

One thing is constant: Always and forever, we would like to thank our wives and life partners, Robin and Lisa, whose love and support make everything possible in our lives. We dedicate this book to them.

INDEX